DESIGNERS' GUIDE TO EUROCODE 3: DESIGN OF STEEL BUILDINGS

Second edition

Eurocode Designers' Guide series

Designers' Guide to EN 1990 Eurocode: Basis of structural design. H. Gulvanessian, J.-A. Calgaro and M. Holický. 978-0-7277-3011-4. Published 2002 [New edition forthcoming: 2012].

Designers' Guide to Eurocode 8: Design of structures for earthquake resistance. EN 1998-1 and EN 1998-5. General rules, seismic actions, design rules for buildings, foundations and retaining structures. M. Fardis, E. Carvalho, A. Elnashai, E. Faccioli, P. Pinto and A. Plumier. 978-0-7277-3348-1. Published 2005.

Designers' Guide to EN 1994-1-1. Eurocode 4: Design of Composite Steel and Concrete Structures, Part 1-1: General Rules and Rules for Buildings. R.P. Johnson and D. Anderson. 978-0-7277-3151-7. Published 2004.

Designers' Guide to Eurocode 7: Geotechnical design. EN 1997-1 General rules. R. Frank, C. Bauduin, R. Driscoll, M. Kavvadas, N. Krebs Ovesen, T. Orr and B. Schuppener. 978-0-7277-3154-8. Published 2004.

Designers' Guide to Eurocode 3: Design of Steel Buildings. EN 1993-1-1, -1-3 and -1-8. Second edition. L. Gardner and D. Nethercot. 978-0-7277-4172-1. Published 2011.

Designers' Guide to Eurocode 2: Design of Concrete Structures. EN 1992-1-1 and EN 1992-1-2 General rules and rules for buildings and structural fire design. R.S. Narayanan and A.W. Beeby. 978-0-7277-3105-0. Published 2005.

Designers' Guide to EN 1994-2. Eurocode 4: Design of composite steel and concrete structures. Part 2 General rules for bridges. C.R. Hendy and R.P. Johnson. 978-0-7277-3161-6. Published 2006 [New edition forthcoming: 2011].

Designers' Guide to EN 1992-2. Eurocode 2: Design of concrete structures. Part 2: Concrete bridges. C.R. Hendy and D.A. Smith. 978-0-7277-3159-3. Published 2007.

Designers' Guide to EN 1991-1-2, EN 1992-1-2, EN 1993-1-2 and EN 1994-1-2. T. Lennon, D.B. Moore, Y.C. Wang and C.G. Bailey. 978-0-7277-3157-9. Published 2007.

Designers' Guide to EN 1993-2. Eurocode 3: Design of steel structures. Part 2: Steel bridges. C.R. Hendy and C.J. Murphy. 978-0-7277-3160-9. Published 2007.

Designers' Guide to EN 1991-1.4. Eurocode 1: Actions on structures, general actions. Part 1-4 Wind actions. N. Cook. 978-0-7277-3152-4. Published 2007.

Designers' Guide to Eurocode 1: Actions on buildings. EN 1991-1-1 and -1-3 to -1-7. H. Gulvanessian, P. Formichi and J.-A. Calgaro. 978-0-7277-3156-2. Published 2009.

Designers' Guide to Eurocode 1: Actions on bridges. EN 1991-1-1, -1-3 to -1-7 and EN 1991-2. J.-A. Calgaro, M. Tschumi and H. Gulvanessian. 978-0-7277-3158-6. Published 2010.

Designers' Guide to Eurocode 6: Design of masonry structures. EN 1996-1-1. J. Morton. 978-0-7277-3155-5. Forthcoming: 2011.

Designers' Guide to Eurocode 5: Design of timber buildings. EN 1995-1-1. A.J. Porteous and P. Ross. 978-0-7277-3162-3. Forthcoming: 2011.

Designers' Guide to Eurocode 8: Design of bridges for earthquake resistance. EN 1998-2 and -5. M.N. Fardis, B. Kolias and A. Pecker. Forthcoming: 2012.

www.icevirtuallibrary.com
www.eurocodes.co.uk

DESIGNERS' GUIDES TO THE EUROCODES

DESIGNERS' GUIDE TO EUROCODE 3: DESIGN OF STEEL BUILDINGS
EN 1993-1-1, -1-3 and -1-8

Second edition

LEROY GARDNER and DAVID A. NETHERCOT
Imperial College London, UK

Series editor
Haig Gulvanessian

Published by ICE Publishing, 40 Marsh Wall, London E14 9TP

Full details of ICE Publishing sales representatives and distributors can be found at:
www.icevirtuallibrary.com/info/printbooksales

First published 2005
Second edition 2011

Eurocodes Expert

Structural Eurocodes offer the opportunity of harmonised design standards for the European construction market and the rest of the world. To achieve this, the construction industry needs to become acquainted with the Eurocodes so that the maximum advantage can be taken of these opportunities.

Eurocodes Expert is a new ICE and Thomas Telford initiative set up to assist in creating a greater awareness of the impact and implementation of the Eurocodes within the UK construction industry.

Eurocodes Expert provides a range of products and services to aid and support the transition to Eurocodes. For comprehensive and useful information on the adoption of the Eurocodes and their implementation process please visit our website on or email eurocodes@thomastelford.com

www.icevirtuallibrary.com

A catalogue record for this book is available from the British Library

ISBN 978-0-7277-4172-1

Typeset by Academic + Technical, Bristol
Index created by Indexing Specialists (UK) Ltd, Hove, East Sussex
Printed and bound by CPI Group (UK) Ltd, Croydon, CR0 4YY

Preface

Now that the UK has adopted the set of structural Eurocodes it is timely to produce revised versions of the series of guides based on their technical content. For the design of steel structures, *Eurocode 3: Design of Steel Structures*, Part 1.1: *General Rules and Rules for Buildings* (EN 1993-1-1), together with its National Annex, is the master document. It is, however, complemented by several other parts, each of which deals with a particular aspect of the design of structural steelwork.

General

This text concentrates on the main provisions of Part 1.1 of the code, but deals with some aspects of Part 1.3 (cold-formed sections), Part 1.5 (plated structures) and Part 1.8 (connections). It does this by presenting and discussing the more important technical provisions, often by making specific reference to actual sections of the code documents. In addition, it makes comparisons with the equivalent provisions in BS 5950, and illustrates the application of certain of the design procedures with a series of worked examples. When dealing with loads and load combinations it makes appropriate reference to the companion Eurocodes EN 1990 and EN 1991.

Layout of this guide

The majority of the text relates to the most commonly encountered design situations. Thus, the procedures for design at the cross-sectional, member and frame level for various situations are covered in some detail. Chapters 1–11 directly reflect the arrangement of the code (i.e. section numbers and equation numbers match those in EN 1993-1-1), and it is for this reason that the chapters vary greatly in length. Guidance on design for the ultimate limit state dominates Part 1.1; this is mirrored herein. In the case of Chapters 12–14, the section numbering does not match the code, and the arrangement adopted is explained at the start of each of these chapters.

All cross-references in this guide to sections, clauses, subclauses, paragraphs, annexes, figures, tables and expressions of EN 1993-1-1 are in *italic type*, which is also used where text from EN 1993-1-1 has been directly reproduced (conversely, quotations from other sources, including other Eurocodes, and cross-references to sections, etc., of this guide, are in roman type). Expressions repeated from EN 1993-1-1 retain their numbering; other expressions have numbers prefixed by D (for Designers' Guide), e.g. equation (D5.1) in Chapter 5.

The Eurocode format specifically precludes reproduction of material from one part to another. The 'basic rules' of the EN 1993-1-1 therefore provide insufficient coverage for the complete design of a structure (e.g. Part 1.1 contains no material on connections, all of which is given in Part 1.8). Thus, in practice, designers will need to consult several parts of the code. It is for this reason that we have elected to base the content of the book on more than just Part 1.1. Readers will also find several references to the UK National Annex. The National Annex provides specific limitations and guidance on the use of a number of provisions. Since these overrule the basic clauses for application in the UK, their use has been included throughout this text. Where appropriate, reference has also been made to sources of non-contradictory complementary information (NCCI).

Acknowledgements

In preparing this text the authors have benefited enormously from discussions and advice from many individuals and groups involved with the Eurocode operation. To each of these we accord our thanks. We are particularly grateful to Charles King of the SCI, who has provided expert advice on many technical matters throughout the production of the book.

L. Gardner
D. A. Nethercot

Contents

Designers' Guide to Eurocode 3: Design of Steel Buildings, 2nd ed.
ISBN 978-0-7277-4172-1

doi: 10.1680/dsb.41721.001

Introduction

The material in this introduction relates to the foreword to the European Standard EN 1993-1-1, *Eurocode 3: Design of Steel Structures*, Part 1.1: *General Rules and Rules for Buildings*. The following aspects are covered:

- Background to the Eurocode programme
- Status and field of application of Eurocodes
- National standards implementing Eurocodes
- Links between Eurocodes and product-harmonised technical specifications (ENs and ETAs)
- Additional information specific to EN 1993-1
- UK National Annex for EN 1993-1-1.

Background to the Eurocode programme

Work began on the set of structural Eurocodes in 1975. For structural steelwork, the responsible committee, under the chairmanship of Professor Patrick Dowling of Imperial College London, had the benefit of the earlier *European Recommendations for the Design of Structural Steelwork*, prepared by the European Convention for Constructional Steelwork in 1978 (ECCS, 1978). Apart from the obvious benefit of bringing together European experts, preparation of this document meant that some commonly accepted design procedures already existed, e.g. the European column curves. Progress was, however, rather slow, and it was not until the mid-1980s that the official draft documents, termed ENVs, started to appear. The original, and unchanged, main grouping of Eurocodes, comprises ten documents: EN 1990, covering the basis of structural design, EN 1991, covering actions on structures, and eight further documents essentially covering each of the structural materials (concrete, steel, masonry, etc.). The full suite of Eurocodes is:

EN 1990 *Eurocode 0: Basis of Structural Design*
EN 1991 *Eurocode 1: Actions on Structures*
EN 1992 *Eurocode 2: Design of Concrete Structures*
EN 1993 *Eurocode 3: Design of Steel Structures*
EN 1994 *Eurocode 4: Design of Composite Steel and Concrete Structures*
EN 1995 *Eurocode 5: Design of Timber Structures*
EN 1996 *Eurocode 6: Design of Masonry Structures*
EN 1997 *Eurocode 7: Geotechnical Design*
EN 1998 *Eurocode 8: Design of Structures for Earthquake Resistance*
EN 1999 *Eurocode 9: Design of Aluminium Structures*

Status and field of application of Eurocodes

Generally, the Eurocodes provide structural design rules that may be applied to complete structures and structural components and other products. Rules are provided for common forms of construction, and it is recommended that specialist advice is sought when considering unusual structures. More specifically, the Eurocodes serve as reference documents that are recognised by the EU member states for the following purposes:

- as a means to prove compliance with the essential requirements of Council Directive 89/106/EEC
- as a basis for specifying contracts for construction or related works
- as a framework for developing harmonised technical specifications for construction products.

National standards implementing Eurocodes

The National Standard implementing Eurocodes (e.g. BS EN 1993-1-1) must comprise the full, unaltered text of that Eurocode, including all annexes (as published by CEN). This may then be preceded by a National Title Page and National Foreword, and, importantly, may be followed by a National Annex.

The National Annex may only include information on those parameters (known as Nationally Determined Parameters (NDPs)) within clauses that have been left open for national choice; these clauses are listed later in this chapter.

Links between Eurocodes and product-harmonised technical specifications (ENs and ETAs)

The clear need for consistency between the harmonised technical specifications for construction products and the technical rules for work is highlighted. In particular, information accompanying such products should clearly state which, if any, NDPs have been taken into account.

Additional information specific to EN 1993-1

As with the Eurocodes for the other structural materials, Eurocode 3 for steel structures is intended to be used in conjunction with EN 1990 and EN 1991, where basic requirements, along with loads (actions) and action combinations are specified. An introduction to the provisions of EN 1990 and EN 1991 may be found in Chapter 14 of this guide. EN 1993-1 is split into 11 parts, listed in Chapter 1 of this guide, each addressing specific steel components, limit states or materials. EN 1993-1 is intended for use by designers and constructors, clients, committees drafting design-related product, testing and execution standards and relevant authorities, and this guide is intended to provide interpretation and guidance on the application of its contents.

UK National Annex for EN 1993-1-1

National choice is allowed in EN 1993-1-1 in the following clauses of the code:

UK National Annex clause	EN 1993-1-1 clause	Comment
NA.2.2	2.3.1(1)	Actions for particular regional or climatic or accidental situations
NA.2.3	3.1(2)	Material properties
NA.2.4	3.2.1(1)	Material properties – use of Table 3.1 or product standards
NA.2.5	3.2.2(1)	Ductility requirements
NA.2.6	3.2.3(1)	Fracture toughness
NA.2.7	3.2.3(3)B	Fracture toughness for buildings
NA.2.8	3.2.4(1)B	Through thickness properties
NA.2.9	5.2.1(3)	Limit on α_{cr} for analysis type
NA.2.10	5.2.2(8)	Scope of application
NA.2.11	5.3.2(3)	Value for relative initial local bow imperfections e_0/L
NA.2.12	5.3.2(11)	Scope of application
NA.2.13	5.3.4(3)	Numerical value for factor k
NA.2.14	6.1(1)B	Numerical values for partial factors γ_{Mi} for buildings
NA.2.15	6.1(1)	Other recommended numerical values for partial factors γ_{Mi}
NA.2.16	6.3.2.2(2)	Imperfection factor α_{LT} for lateral torsional buckling
NA.2.17	6.3.2.3(1)	Numerical values for $\bar{\lambda}_{LT,0}$ and β and geometric limitations for the method
NA.2.18	6.3.2.3(2)	Values for parameter f
NA.2.19	6.3.2.4(1)B	Value for the slenderness limit $\bar{\lambda}_{c0}$
NA.2.20	6.3.2.4(2)B	Value for the modification factor k_{fl}
NA.2.21	6.3.3(5)	Choice between alternative methods 1 and 2 for bending and compression
NA.2.22	6.3.4(1)	Limits of application of general method
NA.2.23	7.2.1(1)B	Vertical deflection limits
NA.2.24	7.2.2(1)B	Horizontal deflection limits
NA.2.25	7.2.3(1)B	Floor vibration limits
NA.2.26	BB.1.3(3)B	Buckling lengths L_{cr}

REFERENCE

ECCS (1978) *European Recommendations for Steel Construction.* European Convention for Constructional Steelwork, Brussels.

Designers' Guide to Eurocode 3: Design of Steel Buildings, 2nd ed.
ISBN 978-0-7277-4172-1

ICE Publishing: All rights reserved
doi: 10.1680/dsb.41721.005

Chapter 1
General

This chapter discusses the general aspects of EN 1993-1-1, as covered in *Section 1* of the code. The following clauses are addressed:

■ Scope	*Clause 1.1*
■ Normative references	*Clause 1.2*
■ Assumptions	*Clause 1.3*
■ Distinction between Principles and Application Rules	*Clause 1.4*
■ Terms and definitions	*Clause 1.5*
■ Symbols	*Clause 1.6*
■ Conventions for member axes	*Clause 1.7*

1.1. Scope

Finalisation of the Eurocodes, the so-called conversion of ENVs into ENs, has seen each of the final documents subdivided into a number of parts, some of which have then been further subdivided. Thus, Eurocode 3 now comprises six parts:

EN 1993-1	*General Rules and Rules for Buildings*
EN 1993-2	*Steel Bridges*
EN 1993-3	*Towers, Masts and Chimneys*
EN 1993-4	*Silos, Tanks and Pipelines*
EN 1993-5	*Piling*
EN 1993-6	*Crane Supporting Structures.*

Part 1 itself consists of 12 sub-parts:

EN 1993-1-1	*General Rules and Rules for Buildings*
EN 1993-1-2	*Structural Fire Design*
EN 1993-1-3	*Cold-formed Members and Sheeting*
EN 1993-1-4	*Stainless Steels*
EN 1993-1-5	*Plated Structural Elements*
EN 1993-1-6	*Strength and Stability of Shell Structures*
EN 1993-1-7	*Strength and Stability of Planar Plated Structures Transversely Loaded*
EN 1993-1-8	*Design of Joints*
EN 1993-1-9	*Fatigue Strength of Steel Structures*
EN 1993-1-10	*Selection of Steel for Fracture Toughness and Through-thickness Properties*
EN 1993-1-11	*Design of Structures with Tension Components Made of Steel*
EN 1993-1-12	*Additional Rules for the Extension of EN 1993 up to Steel Grades S700.*

Part 1.1 of Eurocode 3 is the basic document on which this text concentrates, but designers will need to consult other sub-parts, for example Part 1.8, for information on bolts and welds, and Part 1.10, for guidance on material selection, since no duplication of content is permitted between codes. It is for this reason that it seems likely that designers in the UK will turn first to simplified and more restricted design rules, for example SCI guides and manuals produced by the Institutions of Civil and Structural Engineers, whilst referring to the Eurocode documents themselves when further information is required. Given that some reference to the content of EN 1990 on load combinations and to EN 1991 on loading will also be necessary when conducting design calculations, working directly from the Eurocodes for even the simplest of steel structures requires the simultaneous use of several lengthy documents.

It is worth noting that EN 1993-1-1 is primarily intended for hot-rolled sections with material thickness greater than 3 mm. For cold-formed sections and for material thickness of less than 3 mm, reference should be made to EN 1993-1-3 and to Chapter 13 of this guide. An exception is that cold-formed rectangular and circular hollow sections are also covered by Part 1.1.

Clause numbers in EN 1993-1-1 that are followed by the letter 'B' indicate supplementary rules intended specifically for the design of buildings.

1.2. Normative references

Information on design-related matters is provided in a set of reference standards, of which the most important are:

EN 10025 (in six parts)	*Hot-rolled Steel Products*
EN 10210	*Hot Finished Structured Hollow Sections*
EN 10219	*Cold-formed Structural Hollow Sections*
EN 1090	*Execution of Steel Structures (Fabrication and Erection)*
EN ISO 12944	*Corrosion Protection by Paint Systems.*

1.3. Assumptions

The general assumptions of EN 1990 relate principally to the manner in which the structure is designed, constructed and maintained. Emphasis is given to the need for appropriately qualified designers, appropriately skilled and supervised contractors, suitable materials, and adequate maintenance. Eurocode 3 states that all fabrication and erection should comply with EN 1090.

1.4. Distinction between Principles and Application Rules

EN 1990 explicitly distinguishes between Principles and Application Rules; clause numbers that are followed directly by the letter 'P' are principles, whilst omission of the letter 'P' indicates an application rule. Essentially, Principles are statements for which there is no alternative, whereas Application Rules are generally acceptable methods, which follow the principles and satisfy their requirements. EN 1993-1-1 does not use this notation.

1.5. Terms and definitions

Clause 1.5

Clause 1.5 of EN 1990 contains a useful list of common terms and definitions that are used throughout the structural Eurocodes (EN 1990 to EN 1999). Further terms and definitions specific to EN 1993-1-1 are included in *clause 1.5*. Both sections are worth reviewing because the Eurocodes use a number of terms that may not be familiar to practitioners in the UK.

1.6. Symbols

Clause 1.6

A useful listing of the majority of symbols used in EN 1993-1-1 is provided in *clause 1.6*. Other symbols are defined where they are first introduced in the code. Many of these symbols, especially those with multiple subscripts, will not be familiar to UK designers. However, there is generally good consistency in the use of symbols throughout the Eurocodes, which makes transition between the documents more straightforward.

1.7. Conventions for member axes

The convention for member axes in Eurocode 3 is not the same as that adopted in BS 5950 (where the $x-x$ and $y-y$ axes refer to the major and minor axes of the cross-section respectively. Rather, the Eurocode 3 convention for member axes is as follows:

- $x-x$ along the member
- $y-y$ axis of the cross-section
- $z-z$ axis of the cross-section.

Generally, the $y-y$ axis is the major principal axis (parallel to the flanges), and the $z-z$ axis is the minor principal axis (perpendicular to the flanges). For angle sections, the $y-y$ axis is parallel to the smaller leg, and the $z-z$ axis is perpendicular to the smaller leg. For cross-sections where the major and minor principal axes do not coincide with the $y-y$ and $z-z$ axes, such as for angle sections, then these axes should be referred to as $u-u$ and $v-v$, respectively. The note at the

Clause 1.7

end of *clause 1.7* is important when designing such sections, because it states that '*All rules in*

Figure 1.1. Dimensions and axes of sections in Eurocode 3

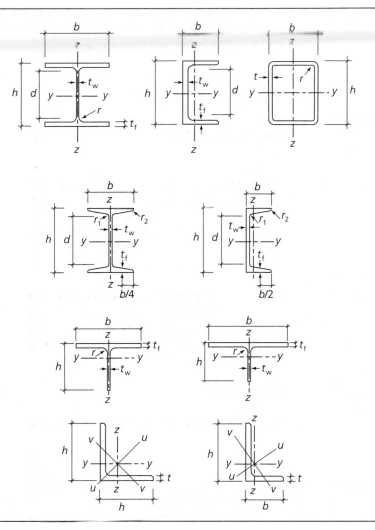

this Eurocode relate to the principal axis properties, which are generally defined by the axes y–y and z–z but for sections such as angles are defined by the axes u–u and v–v' (i.e. for angles and similar sections, the u–u and v–v axes properties should be used in place of the y–y and z–z axes properties).

Figure 1.1 defines the important dimensions and axes for the common types of structural steel cross-section.

Designers' Guide to Eurocode 3: Design of Steel Buildings, 2nd ed.
ISBN 978-0-7277-4172-1

ICE Publishing: All rights reserved
doi: 10.1680/dsb.41721.009

Chapter 2
Basis of design

This chapter discusses the basis of design, as covered in *Section 2* of EN 1993-1-1 and Section 2 of EN 1990. The following clauses are addressed:

- Requirements   *Clause 2.1*
- Principles of limit state design   *Clause 2.2*
- Basic variables   *Clause 2.3*
- Verification by the partial factor method   *Clause 2.4*
- Design assisted by testing   *Clause 2.5*

2.1.   Requirements

The general approach of Eurocode 3 is essentially the same as that of BS 5950, being based on limit state principles using partial safety factors. The approach is set down in detail in EN 1990, with additional explanation to be found in the *Designers' Guide to EN 1990, Eurocode: Basis of Structural Design* (Gulvanessian *et al.*, 2002). Chapter 14 of this guide gives some introductory recommendations on the use of EN 1990 and EN 1991, including the specification of loading and the development of load combinations. Further references to EN 1990 are made throughout the guide.

The basic requirements of EN 1990 state that a structure shall be designed to have adequate:

- structural resistance
- serviceability
- durability
- fire resistance (for a required period of time)
- robustness (to avoid disproportionate collapse due to damage from events such as explosion, impact and consequences of human error).

Clause 2.1.1(2) states that these '*basic requirements shall be deemed to be satisfied where limit state design is used in conjunction with the partial factor method and the load combinations given in EN 1990 together with the actions given in EN 1991*'.

Clause 2.1.1(2)

Outline notes on the design working life, durability and robustness of steel structures are given in *clause 2.1.3*. Design working life is defined in Section 1 of EN 1990 as the 'assumed period for which a structure or part of it is to be used for its intended purpose with anticipated maintenance but without major repair being necessary'. The design working life of a structure will generally be determined by its application (and may be specified by the client). Indicative design working lives are given in Table 2.1 (Table 2.1 of EN 1990), which may be useful, for example, when considering time-dependent effects such as fatigue and corrosion.

Clause 2.1.3

Durability is discussed in more detail in Chapter 4 of this guide, but the general guidance of *clause 2.1.3.1* explains that steel structures should be designed (protected) against corrosion, detailed for sufficient fatigue life, designed for wearing, designed for accidental actions, and inspected and maintained at appropriate intervals (with consideration given in the design to ensure that parts susceptible to these effects are easily accessible).

Clause 2.1.3.1

2.2.   Principles of limit state design

General principles of limit state design are set out in Section 3 of EN 1990. *Clause 2.2* reminds the designer of the importance of ductility. It states that the cross-section and member resistance

Clause 2.2

Table 2.1. Indicative design working life

Design working life category	Indicative design working life (years)	Examples
1	10	Temporary structures (not those that can be dismantled with a view to being reused)
2	10–25	Replaceable structural parts, e.g. gantry girders and bearings
3	15–30	Agricultural and similar structures
4	50	Building structures and other common structures
5	100	Monumental building structures, bridges and other civil engineering structures

models given in Eurocode 3 assume that the material displays sufficient ductility. In order to ensure that these material requirements are met, reference should be made to *Section 3* (and Chapter 3 of this guide).

2.3. Basic variables

General information regarding basic variables is set out in Section 4 of EN 1990. Loads, referred to as actions in the structural Eurocodes, should be taken from EN 1991, whilst partial factors and the combination of actions are covered in EN 1990. Some preliminary guidance on actions and their combination is given in Chapter 14 of this guide.

2.4. Verification by the partial factor method

Throughout EN 1993-1-1, material properties and geometrical data are required in order to calculate the resistance of structural cross-sections and members. The basic equation governing the resistance of steel structures is given by *equation (2.1)*:

$$R_\mathrm{d} = \frac{R_\mathrm{k}}{\gamma_\mathrm{M}} \qquad (2.1)$$

where R_d is the design resistance, R_k is the characteristic resistance and γ_M is a partial factor which accounts for material, geometric and modelling uncertainties (and is the product of γ_m and γ_Rd).

However, for practical design purposes, and to avoid any confusion that may arise from terms such as 'nominal values', 'characteristic values' and 'design values', the following guidance is provided:

Clause NA.2.4

- For material properties, the nominal values given in Table 3.1 of this guide may be used (as characteristic values) for design. These values have been taken, as advised in *clause NA.2.4* of the UK National Annex, as the minimum specified values from product standards, such as EN 10025 and EN 10210.

Clause 2.4.2(1)
Clause 2.4.2(2)

- For cross-section and system geometry, dimensions may be taken from product standards or drawings for the execution of the structure to EN 1090 and treated as nominal values – these values may also be used in design (*clause 2.4.2(1)*).
- *Clause 2.4.2(2)* highlights that the design values of geometric imperfections, used primarily for structural analysis and member design (see Section 5), are equivalent geometric imperfections that take account of actual geometric imperfections (e.g. initial out-of-straightness), structural imperfections due to fabrication and erection (e.g. misalignment), residual stresses and variation in yield strength throughout the structural component.

2.5. Design assisted by testing

Clause 2.5

An important feature of steel design in the UK is the reliance on manufacturers' design information for many products, such as purlins and metal decking. *Clause 2.5* authorises this process, with the necessary detail being given in Annex D of EN 1990.

REFERENCE

Gulvanessian H, Calgaro J-A and Holický M (2002) *Designers' Guide to EN 1990, Eurocode: Basis of Structural Design*. Thomas Telford, London.

Designers' Guide to Eurocode 3: Design of Steel Buildings, 2nd ed.
ISBN 978-0-7277-4172-1

ICE Publishing: All rights reserved
doi: 10.1680/dsb.41721.011

Chapter 3
Materials

This chapter is concerned with the guidance given in EN 1993-1-1 for materials, as covered in *Section 3* of the code. The following clauses are addressed:

- General   *Clause 3.1*
- Structural steel   *Clause 3.2*
- Connecting devices   *Clause 3.3*
- Other prefabricated products in buildings   *Clause 3.4*

3.1. General

Clause NA.2.4 of the UK National Annex states that nominal values of material properties should be taken from the relevant product standard. These values may then be used in the design expressions given throughout the code.

Clause NA.2.4

3.2. Structural steel

As noted above, *clause NA.2.4* of the UK National Annex directs designers to the product standards for the determination of material properties. The key standards are EN 10025-2 for hot-rolled flat and long products (including I and H sections) and EN 10210-1 for hot-finished structural hollow sections. Values for both yield strength f_y (taken as R_{eH} from the product standards) and ultimate tensile strength f_u (taken as the lower value of the range of R_m given in the product standards) are presented in Table 3.1. Although not explicitly stated in EN 1993-1-1, it is recommended that, for rolled sections, the thickness of the thickest element is used to define a single yield strength to be applied to the entire cross-section.

Clause NA.2.4

In order to ensure structures are designed to EN 1993-1-1 with steels that possess adequate ductility, the following requirements are set out in *clause NA.2.5* of the UK National Annex.

Clause NA.2.5

For elastic analysis:

- $f_u/f_y \geq 1.10$
- elongation at failure >15% (based on a gauge length of $5.65\sqrt{A_0}$, where A_0 is the original cross-sectional area)
- $\varepsilon_u \geq 15\varepsilon_y$, where ε_u is the ultimate strain and ε_y is the yield strain.

For plastic analysis:

- $f_u/f_y \geq 1.15$
- elongation at failure >15% (based on a gauge length of $5.65\sqrt{A_0}$)
- $\varepsilon_u \geq 20\varepsilon_y$.

All steel grades listed in Table 3.1 meet these criteria, so do not have to be explicitly checked. In general, it is only the higher-strength grades that may fail to meet the ductility requirements.

In order to avoid brittle fracture, materials need sufficient fracture toughness at the lowest service temperature expected to occur within the intended design life of the structure. In the UK, the lowest service temperature should normally be taken as −5°C for internal steelwork and −15°C for external steelwork, as stated in *clause NA.2.6* of the UK National Annex. Fracture toughness and design against brittle fracture is covered in detail in Eurocode 3 – Part 1.10.

Clause NA.2.6

Table 3.1. Values for yield strength f_y and ultimate tensile strength f_u from product standards (EN 10025-2 and EN 10210-1)

Steel grade	Thickness range (mm)	Yield strength, f_y (N/mm^2)	Thickness range (mm)	Ultimate tensile strength, f_u (N/mm^2)
S235	$t \leq 16$	235	$t < 3$	360
	$16 < t \leq 40$	225		
	$40 < t \leq 63$	215	$3 \leq t \leq 100$	360
	$63 < t \leq 80$	215		
	$80 < t \leq 100$	215		
S275	$t \leq 16$	275	$t < 3$	430
	$16 < t \leq 40$	265		
	$40 < t \leq 63$	255	$3 \leq t \leq 100$	410
	$63 < t \leq 80$	245		
	$80 < t \leq 100$	235		
S355	$t \leq 16$	355	$t < 3$	510
	$16 < t \leq 40$	345		
	$40 < t \leq 63$	335	$3 \leq t \leq 100$	470
	$63 < t \leq 80$	325		
	$80 < t \leq 100$	315		

Clause 3.2.6

Design values of material coefficients to be used in EN 1993-1-1 are given in *clause 3.2.6* as follows:

- modulus of elasticity:

  $E = 210\,000 \ \text{N/mm}^2$

- shear modulus:

  $$G = \frac{E}{2(1 + \nu)} \approx 81\,000 \ \text{N/mm}^2$$

- Poisson's ratio:

  $\nu = 0.3$

- coefficient of thermal expansion:

  $\alpha = 12 \times 10^{-6}/^\circ\text{C}$

  (for temperatures below 100°C).

Those familiar with design to British Standards will notice a marginal (approximately 2%) difference in the value of Young's modulus adopted in EN 1993-1-1, which is 210 000 N/mm^2, compared with 205 000 N/mm^2.

3.3.   Connecting devices

Requirements for fasteners, including bolts, rivets and pins, and for welds and welding consumables are given in Eurocode 3 – Part 1.8, and are discussed in Chapter 12 of this guide.

3.4.   Other prefabricated products in buildings

Clause 3.4(1)B

Clause 3.4(1)B simply notes that any semi-finished or finished structural product used in the structural design of buildings must comply with the relevant EN product standard or ETAG (European Technical Approval Guideline) or ETA (European Technical Approval).

Designers' Guide to Eurocode 3: Design of Steel Buildings, 2nd ed.
ISBN 978-0-7277-4172-1

ICE Publishing: All rights reserved
doi: 10.1680/dsb.41721.013

Chapter 4
Durability

This short chapter concerns the subject of durability and covers the material set out in *Section 4* of EN 1993-1-1, with brief reference to EN 1990.

Durability may be defined as the ability of a structure to remain fit for its intended or foreseen use throughout its design working life, with an appropriate level of maintenance.

For basic durability requirements, Eurocode 3 directs the designer to Section 2.4 of EN 1990, where it is stated that 'the structure shall be designed such that deterioration over its design working life does not impair the performance of the structure below that intended, having due regard to its environment and the anticipated level of maintenance'.

The following factors are included in EN 1990 as ones that should be taken into account in order to achieve an adequately durable structure:

- the intended or foreseeable use of the structure
- the required design criteria
- the expected environmental conditions
- the composition, properties and performance of the materials and products
- the properties of the soil
- the choice of the structural system
- the shape of members and structural detailing
- the quality of workmanship and level of control
- the particular protective measures
- the intended maintenance during the design working life.

A more detailed explanation of the basic Eurocode requirements for durability has been given by Gulvanessian *et al.* (2002), and a general coverage of the subject of durability in steel (bridge) structures is available (Corus, 2002).

Of particular importance for steel structures are the effects of corrosion, mechanical wear and fatigue. Therefore, parts susceptible to these effects should be easily accessible for inspection and maintenance.

In buildings, a fatigue assessment is not generally required. However, EN 1993-1-1 highlights several cases where fatigue should be considered, including where cranes or vibrating machinery are present, or where members may be subjected to wind- or crowd-induced vibrations.

Corrosion would generally be regarded as the most critical factor affecting the durability of steel structures, and the majority of points listed above influence the matter. Particular consideration has to be given to the environmental conditions, the intended maintenance schedule, the shape of members and structural detailing, the corrosion protection measures, and the material composition and properties. For aggressive environments, such as coastal sites, and where elements cannot be easily inspected, extra attention is required. Corrosion protection does not need to be applied to internal building structures, if the internal relative humidity does not exceed 80%.

In addition to suitable material choice, a designer can significantly influence the durability of the steel structure through good detailing. Poor (left-hand column) and good (right-hand

Figure 4.1. Poor and good design features for durability (Baddoo and Burgan, 2001)

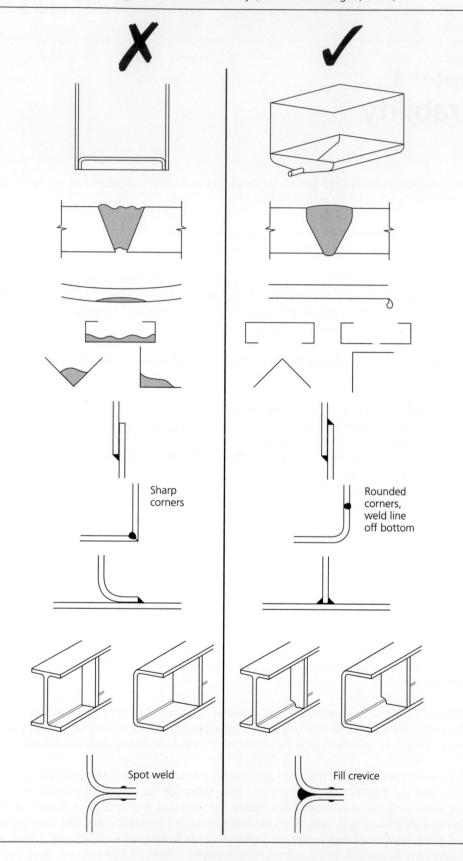

column) design features are shown in Figure 4.1. Additionally, corrosion cannot take place without the presence of an electrolyte (e.g. water) – suitable drainage and good thermal insulation to prevent cold-bridging (leading to condensation) are therefore of key importance.

REFERENCES

Baddoo NR and Burgan BA (2001) *Structural Design of Stainless Steel*. Steel Construction Institute, Ascot, P291.

Corus (2002) *Corrosion Protection of Steel Bridges*. Corus Construction Centre, Scunthorpe.

Gulvanessian H, Calgaro J-A and Holický M (2002) *Designers' Guide to EN 1990 Eurocode: Basis of Structural Design*. Thomas Telford, London.

Designers' Guide to Eurocode 3: Design of Steel Buildings, 2nd ed.
ISBN 978-0-7277-4172-1

ICE Publishing: All rights reserved
doi: 10.1680/dsb.41721.017

Chapter 5
Structural analysis

This chapter concerns the subject of structural analysis and classification of cross-sections for steel structures. The material in this chapter is covered in *Section 5* of EN 1993-1-1, and the following clauses are addressed:

- Structural modelling for analysis   *Clause 5.1*
- Global analysis   *Clause 5.2*
- Imperfections   *Clause 5.3*
- Methods of analysis considering material non-linearities   *Clause 5.4*
- Classification of cross-sections   *Clause 5.5*
- Cross-section requirements for plastic global analysis   *Clause 5.6*

Before the strength of cross-sections and the stability of members can be checked against the requirements of the code, the internal (member) forces and moments within the structure need to be determined from a global analysis. Four distinct types of global analysis are possible:

1. first-order elastic – initial geometry and fully linear material behaviour
2. second-order elastic – deformed geometry and fully linear material behaviour
3. first-order plastic – initial geometry and non-linear material behaviour
4. second-order plastic – deformed geometry and non-linear material behaviour.

Typical predictions of load–deformation response for the four types of analysis are shown in Figure 5.1.

Clause 5.2 explains how a second-order analysis (i.e. one in which the effect of deformations significantly altering the member forces or moments or the structural behaviour is explicitly allowed for) should be conducted. *Clause 5.3* deals with the inclusion of geometrical imperfections both for the overall structure and for individual members, whilst *clause 5.4* covers the inclusion of material non-linearity (i.e. plasticity) in the various types of analysis.

Clause 5.2

Clause 5.3
Clause 5.4

5.1. Structural modelling for analysis

Clause 5.1 outlines the fundamentals and basic assumptions relating to the modelling of structures and joints. It states that the chosen (calculation) model must be appropriate and must accurately reflect the structural behaviour for the limit state under consideration. In general, an elastic global analysis would be used when the performance of the structure is governed by serviceability criteria.

Clause 5.1

Elastic analysis is also routinely used to obtain member forces for subsequent use in the member checks based on the ultimate strength considerations of *Section 6*. This is well accepted, can be shown to lead to safe solutions and has the great advantage that superposition of results may be used when considering different load cases. For certain types of structure, e.g. portal frames, a plastic hinge form of global analysis may be appropriate; very occasionally, for checks on complex or particularly sensitive configurations, a full material and geometrical non-linear approach may be required.

The choice between a first- and a second-order analysis should be based upon the flexibility of the structure; in particular, the extent to which ignoring second-order effects might lead to an unsafe approach due to underestimation of some of the internal forces and moments.

Figure 5.1. Prediction of load–deformation response from structural analysis

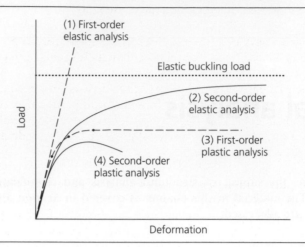

Eurocode 3 recognises the same three types of joint, in terms of their effect on the behaviour of the frame structure, as BS 5950: Part 1. However, the Eurocode uses the term 'semi-continuous' for behaviour between 'simple' and 'continuous', and covers this form of construction in Part 1.8. Consideration of this form of construction and the design of connections in general is covered in Chapter 12 of this guide. Examples of beam-to-column joints that exhibit nominally simple, semi-continuous and continuous behaviour are shown in Figure 5.2.

5.2. Global analysis
5.2.1 Effects of deformed geometry on the structure

Clause 5.2.1

Guidance on the choice between using a first- or second-order global analysis is given in *clause 5.2.1*. The clause states that a first-order analysis may be used provided that the effects of deformations (on the internal member forces or moments and on the structural behaviour) are negligible. This may be assumed to be the case provided that *equation (5.1)* is satisfied:

$$\alpha_{cr} \geq 10 \qquad \text{for elastic analysis}$$
$$\alpha_{cr} \geq 15 \qquad \text{for plastic analysis}$$

(5.1)

where the parameter α_{cr} is the ratio of the elastic critical buckling load for global instability of the structure F_{cr} to the design loading on the structure F_{Ed}, as given by equation (D5.1).

$$\alpha_{cr} = \frac{F_{cr}}{F_{Ed}}$$

(D5.1)

Clause NA.2.9

For plastic analysis of clad structures, provided that the stiffening effect of masonry infill wall panels or diaphragms of profiled steel sheeting are not taken into account, *clause NA.2.9* of the UK National Annex allows second-order effects to be ignored to a lower limit of $\alpha_{cr} \geq 10$. For plastic analysis of portal frames subject to gravity loads only (but with frame imperfections

Figure 5.2. Typical beam-to-column joints. (a) Simple joint. (b) Semi-continuous joint. (c) Rigid joint

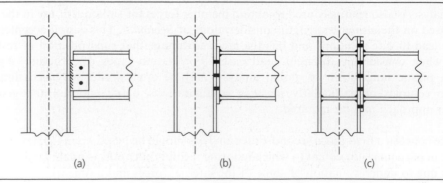

or equivalent horizontal forces), this limit is lowered further to $\alpha_{cr} \geq 5$, provided the conditions set out in *clause NA.2.9* of the UK National Annex are met.

Clause NA.2.9

Essentially, the designer is faced with two questions: is a second-order approach necessary? And if so, how should it be conducted? Guidance on both matters is provided in *clauses 5.2.1* and *5.2.2*.

Clause 5.2.1
Clause 5.2.2

In many cases, experienced engineers will 'know' that a first-order approach will be satisfactory for the form of structure under consideration. In case of doubt, the check (against *equation (5.1)*) should, of course, be made explicitly. Increasingly, standard, commercially available software that includes a linear elastic frame analysis capability will also provide an option to calculate the elastic critical load F_{cr} for the frame.

As an alternative, for portal frames (with shallow roof slopes of less than 26°) and beam and column plane frames, for the important sway mode (the form of instability that in most cases is likely to be associated with the lowest value of F_{cr} and is therefore likely to be the controlling influence on the need, or otherwise, for a second-order treatment), *equation (5.2)* provides an explicit means for determining α_{cr} using only frame geometry, the applied loads and a first-order determined sway displacement:

$$\alpha_{cr} = \left(\frac{H_{Ed}}{V_{Ed}}\right)\left(\frac{h}{\delta_{H,Ed}}\right) \qquad (5.2)$$

where

H_{Ed} is the horizontal reaction at the bottom of the storey due to the horizontal loads (e.g. wind) and the fictitious horizontal loads

V_{Ed} is the total design vertical load on the structure at the level of the bottom of the storey under consideration

$\delta_{H,Ed}$ is the horizontal deflection at the top of the storey under consideration relative to the bottom of the storey, with all horizontal loads (including the fictitious loads) applied to the structure

h is the storey height.

Note that NCCI SN004 (SCI, 2005) allows the calculation of α_{cr} through *equation (5.2)* to be based on the fictitious horizontal loads and corresponding deflections only.

Resistance to sway deformations can be achieved by a variety of means, e.g. a diagonal bracing system (Figure 5.3), rigid connections or a concrete core. In many cases, a combination of

Figure 5.3. External diagonal bracing system (Sanomatalo Building, Helsinki)

Figure 5.4. Swiss Re building, London

systems may be employed, for example the Swiss Re building in London (Figure 5.4) utilises a concrete core plus a perimeter grid of diagonally interlocking steel elements.

For regular multi-storey frames, α_{cr} should be calculated for each storey, although it is the base storey that will normally control. *Equation (5.1)* must be satisfied for each storey for a first-order analysis to suffice. When using *equation (5.2)* it is also necessary that the axial compressive forces in individual members meet the restriction of *clause 5.2.1(4)*.

Clause 5.2.1(4)

5.2.2 Structural stability of frames

Clause 5.2.2

Although it is possible, as is stated in *clause 5.2.2*, to allow for all forms of geometrical and material imperfections in a second-order global analysis, such an approach requires specialist software and is only likely to be used very occasionally in practice, at least for the foreseeable future. A much more pragmatic treatment separates the effects and considers global (i.e. frame imperfections) in the global analysis and local (i.e. member imperfections) in the member checks. Thus option (b) of *clause 5.2.2(4)* will be the most likely choice. Software is now available commercially that will conduct true second-order analysis as described in *clause 5.2.2(4)*. Output from such programs gives the enhanced member forces and moments directly; they can then be used with the member checks of *clause 6.3*. Alternatively, it may be possible to enhance the moments and forces calculated by a linear analysis so as to approximate the second-order values using *clauses 5.2.2(5)* and *5.2.2(6)*. This approach is commonly referred to as the amplified sway method, with the amplification factor k_r defined in *clause NA.2.10* of the UK National Annex. As a further alternative, the method of 'substitutive members' is also permitted. This requires the determination of a 'buckling length' for each member, ideally extracted from the results of a global buckling analysis, i.e. the method used to determine F_{cr} for the frame. Conceptually, it is equivalent to the well-known effective length approach used in conjunction with an interaction formula, in which an approximation to the effect of the enhanced moments within the frame is made by using a reduced axial resistance for the

Clause 5.2.2(4)

Clause 6.3
Clause 5.2.2(5)
Clause 5.2.2(6)
Clause NA.2.10

Figure 5.5. Replacement of initial imperfections by equivalent (fictitious) horizontal forces

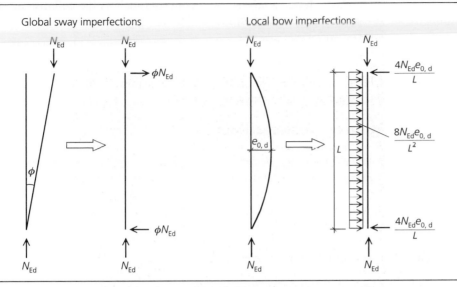

compression members based on considerations of their conditions of restraint. Whilst this approach may be shown to be reasonable for relatively simple, standard cases, it becomes increasingly less accurate as the complexity of the arrangement being considered increases.

5.3. Imperfections

Account should be taken of two types of imperfection:

- global imperfections for frames and bracing systems
- local imperfections for members.

The former require explicit consideration in the overall structural analysis; the latter can be included in the global analysis, but will usually be treated implicitly within the procedures for checking individual members.

Details of the exact ways in which global imperfections should be included are provided in *clauses 5.3.2* and *5.3.3* for frames and bracing systems respectively. Essentially, one of two approaches may be used:

Clause 5.3.2
Clause 5.3.3

- defining the geometry of the structure so that it accords with the imperfect shape, e.g. allowing for an initial out-of-plumb when specifying the coordinates of the frame
- representing the effects of the geometrical imperfections by a closed system of equivalent fictitious forces (replacement of initial imperfections by equivalent horizontal forces is shown in Figure 5.5).

For the former, it is suggested that the initial shape be based on the mode shape associated with the lowest elastic critical buckling load. For the latter, a method to calculate the necessary loads is provided. Imperfection magnitudes for both global sway imperfections (for frames) and local bow imperfections (for members) are defined in *clause 5.3.2(3)* and *clause NA.2.11* of the UK National Annex.

Clause 5.3.2(3)
Clause NA.2.11

5.4. Methods of analysis considering material non-linearities

This section sets out in rather more detail than is customary in codes the basis on which the pattern of the internal forces and moments in a structure necessary for the checking of individual member resistances should be calculated. Thus, *clause 5.4.2* permits the use of linear elastic analysis, including use in combination with member checks on an ultimate strength basis. *Clause 5.4.3* distinguishes between three variants of plastic analysis:

Clause 5.4.2

Clause 5.4.3

- elastic–plastic, using plastic hinge theory – likely to be available in only a few specialised pieces of software

- non-linear plastic zone – essentially a research or investigative tool
- rigid–plastic – simple plastic hinge analysis using concepts such as the collapse mechanism; commonly used for portal frames and continuous beams.

Various limitations on the use of each approach are listed. These align closely with UK practice, particularly the restrictions on the use of plastic analysis in terms of the requirement for restraints against out-of-plane deformations, the use of at least singly symmetrical cross-sections and the need for rotation capacity in the plastic hinge regions.

5.5. Classification of cross-sections
5.5.1 Basis

Clause 5.5.1
Clause 6.2

Clause 5.5
Clause 6.2

Determining the resistance (strength) of structural steel components requires the designer to consider firstly the cross-sectional behaviour and secondly the overall member behaviour. *Clauses 5.5.1* and *6.2* cover the cross-sectional aspects of the design process. Whether in the elastic or inelastic material range, cross-sectional resistance and rotation capacity are limited by the effects of local buckling. As in BS 5950, Eurocode 3 accounts for the effects of local buckling through cross-section classification, as described in *clause 5.5*. Cross-sectional resistances may then be determined from *clause 6.2*.

In Eurocode 3, cross-sections are placed into one of four behavioural classes depending upon the material yield strength, the width-to-thickness ratios of the individual compression parts (e.g. webs and flanges) within the cross-section, and the loading arrangement. The classifications from BS 5950 of plastic, compact, semi-compact and slender are replaced in Eurocode 3 with Class 1, Class 2, Class 3 and Class 4, respectively.

5.5.2 Classification of cross-sections
Definition of classes

Clause 5.5.2(1)

The Eurocode 3 definitions of the four classes are as follows (*clause 5.5.2(1)*):

- Class 1 cross-sections are those which can form a plastic hinge with the rotation capacity required from plastic analysis without reduction of the resistance.
- Class 2 cross-sections are those which can develop their plastic moment resistance, but have limited rotation capacity because of local buckling.
- Class 3 cross-sections are those in which the elastically calculated stress in the extreme compression fibre of the steel member assuming an elastic distribution of stresses can reach the yield strength, but local buckling is liable to prevent development of the plastic moment resistance.
- Class 4 cross-sections are those in which local buckling will occur before the attainment of yield stress in one or more parts of the cross-section.

The moment–rotation characteristics of the four classes are shown in Figure 5.6. Class 1 cross-sections are fully effective under pure compression, and are capable of reaching and maintaining their full plastic moment in bending (and may therefore be used in plastic design). Class 2 cross-sections have a somewhat lower deformation capacity, but are also fully effective in pure compression, and are capable of reaching their full plastic moment in bending. Class 3 cross-sections are fully effective in pure compression, but local buckling prevents attainment of the full plastic moment in bending; bending moment resistance is therefore limited to the (elastic) yield moment. For Class 4 cross-sections, local buckling occurs in the elastic range. An effective cross-section is therefore defined based on the width-to-thickness ratios of individual plate elements, and this is used to determine the cross-sectional resistance. In hot-rolled design the majority of standard cross-sections will be Class 1, 2 or 3, where resistances may be based on gross section properties obtained from section tables. Effective width formulations are not contained in Part 1.1 of Eurocode 3, but are instead to be found in Part 1.5; these are discussed later in this section.

For cold-formed cross-sections, which are predominantly of an open nature (e.g. a channel section) and of light-gauge material, design will seldom be based on the gross section properties; the design requirements for cold-formed members are covered in Eurocode 3 – Part 1.3 and in Chapter 14 of this guide.

Figure 5.6. The four behavioural classes of cross-section defined by Eurocode 3

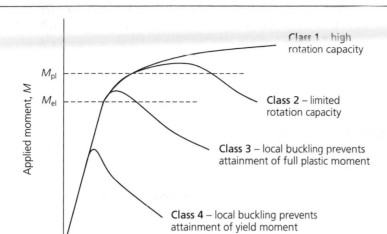

Assessment of individual parts

Each compressed (or partially compressed) element is assessed individually against the limiting width-to-thickness ratios for Class 1, 2 and 3 elements defined in *Table 5.2* (see Table 5.1). An element that fails to meet the Class 3 limits should be taken as Class 4. *Table 5.2* contains three sheets. Sheet 1 is for internal compression parts, defined as those supported along each edge by an adjoining flange or web. Sheet 2 is for outstand flanges, where one edge of the part is supported by an adjoining flange or web and the other end is free. Sheet 3 deals with angles and tubular (circular hollow) sections.

The limiting width-to-thickness ratios are modified by a factor ε that is dependent upon the material yield strength. (For circular hollow members the diameter-to-thickness ratios are modified by ε^2.) ε is defined as

$$\varepsilon = \sqrt{235/f_y} \tag{D5.2}$$

where f_y is the nominal yield strength of the steel as defined in Table 3.1. Clearly, increasing the nominal material yield strength results in stricter classification limits. It is worth noting that the definition of ε in Eurocode 3 (equation (D5.2)) utilises a base value of 235 N/mm², simply because grade S235 steel is widely used and regarded as the normal grade throughout Europe. In comparison, BS 5950 and BS 5400 use 275 and 355 N/mm² as base values, respectively.

The nominal yield strength depends upon the steel grade, the standard to which the steel is produced, and the nominal thickness of the steel element under consideration. The UK National Annex specifies that material properties are taken from the relevant product standard, as described in Section 3.2 of this guide – values have been extracted from the product standards and included in Table 3.1 of this guide.

The classification limits provided in *Table 5.2* assume that the cross-section is stressed to yield, although where this is not the case, *clauses 5.5.2(9)* and *5.5.2(10)* may allow some relaxation of the Class 3 limits. For cross-sectional checks and when buckling resistances are determined by means of a second-order analysis, using the member imperfections of *clause 5.3*, Class 4 cross-sections may be treated as Class 3 if the width-to-thickness ratios are less than the limiting proportions for Class 3 sections when ε is increased by a factor to give the definition of equation (D5.3):

$$\varepsilon = \sqrt{235/f_y} \sqrt{\frac{f_y/\gamma_{M0}}{\sigma_{com,Ed}}} \tag{D5.3}$$

Clause 5.5.2(9)
Clause 5.5.2(10)

Clause 5.3

Table 5.1 (sheet 1 of 3). Maximum width-to-thickness ratios for compression parts (*Table 5.2* of EN 1993-1-1)

Internal compression parts			
Class	Part subject to bending	Part subject to compression	Part subject to bending and compression
Stress distribution in parts (compression positive)			
1	$c/t \le 72\varepsilon$	$c/t \le 33\varepsilon$	when $\alpha > 0.5$: $c/t \le \dfrac{396\varepsilon}{13\alpha - 1}$ when $\alpha \le 0.5$: $c/t \le \dfrac{36\varepsilon}{\alpha}$
2	$c/t \le 83\varepsilon$	$c/t \le 38\varepsilon$	when $\alpha > 0.5$: $c/t \le \dfrac{456\varepsilon}{13\alpha - 1}$ when $\alpha \le 0.5$: $c/t \le \dfrac{41.5\varepsilon}{\alpha}$
Stress distribution in parts (compression positive)			
3	$c/t \le 124\varepsilon$	$c/t \le 42\varepsilon$	when $\psi > -1$: $c/t \le \dfrac{42\varepsilon}{0.67 + 0.33\psi}$ when $\psi \le -1^{*}$): $c/t \le 62\varepsilon(1 - \psi)$

$\varepsilon = \sqrt{235/f_y}$	f_y	235	275	355	420	460
	ε	1.00	0.92	0.81	0.75	0.71

*) $\psi \le -1$ applies where either the compression stress $\sigma < f_y$ or the tensile strain $\varepsilon_y > f_y/E$

where $\sigma_{\mathrm{com,Ed}}$ should be taken as the maximum design compressive stress that occurs in the member.

Clause 6.3

For conventional member design, whereby buckling resistances are determined using the buckling curves defined in *clause 6.3*, no modification to the basic definition of ε (given by equation (D5.2)) is permitted, and the limiting proportions from *Table 5.2* should always be applied.

Notes on *Table 5.2* of EN 1993-1-1

The purpose of this subsection is to provide notes of clarification on *Table 5.2* (reproduced here as Table 5.1).

Table 5.1 (sheet 2 of 3). Maximum width-to-thickness ratios for compression parts (*Table 5.2* of EN 1993-1-1)

Outstand flanges

Class	Part subject to compression	Part subject to bending and compression	
		Tip in compression	Tip in tension
Stress distribution in parts (compression positive)			
1	$c/t \leq 9\varepsilon$	$c/t \leq \dfrac{9\varepsilon}{\alpha}$	$c/t \leq \dfrac{9\varepsilon}{\alpha\sqrt{\alpha}}$
2	$c/t \leq 10\varepsilon$	$c/t \leq \dfrac{10\varepsilon}{\alpha}$	$c/t \leq \dfrac{10\varepsilon}{\alpha\sqrt{\alpha}}$
Stress distribution in parts (compression positive)			
3	$c/t \leq 14\varepsilon$	$c/t \leq 21\varepsilon\sqrt{k_\sigma}$ For k_σ see EN 1993-1-5	

$\varepsilon = \sqrt{235/f_y}$	f_y	235	275	355	420	460
	ε	1.00	0.92	0.81	0.75	0.71

The following points are worth noting:

1. For sheets 1 and 2 of *Table 5.2*, all classification limits are compared with c/t ratios (compressive width-to-thickness ratios), with the appropriate dimensions for c and t taken from the accompanying diagrams. In this guide, c_f and c_w are used to distinguish between flange and web compressed widths, respectively.
2. The compression widths c defined in sheets 1 and 2 always adopt the dimensions of the flat portions of the cross-sections, i.e. root radii and welds are explicitly excluded from the measurement, as emphasised by Figure 5.7. This was not the case in the ENV version of Eurocode 3 or DS 5950, where generally more convenient measures were adopted (such as for the width of an outstand flange of an I section, taken as half the total flange width).
3. Implementation of point 2 and re-analysis of test results have enabled Eurocode 3 to offer the same classification limits for both rolled and welded cross-sections.
4. For rectangular hollow sections where the value of the internal corner radius is not known, it may be assumed that the compression width c can be taken as equal to $b - 3t$.

The factor k_σ that appears in sheet 2 of *Table 5.2* is a buckling factor, which depends on the stress distribution and boundary conditions in the compression element. Calculation of k_σ is described in Section 6.2.2 of this guide, and should be carried out with reference to Part 1.5 of the code.

Overall cross-section classification

Once the classification of the individual parts of the cross-section is determined, Eurocode 3 allows the overall cross-section classification to be defined in one of two ways:

Table 5.1 (sheet 3 of 3). Maximum width-to-thickness ratios for compression parts (*Table 5.2* of EN 1993-1-1)

Angles		
Refer also to 'Outstand flanges' (see sheet 2 of 3)		Does not apply to angles in continuous contact with other components
Class	Section in compression	
Stress distribution across section (compression positive)		
3	$h/t \leq 15\varepsilon$: $\dfrac{b+h}{2t} \leq 11.5\varepsilon$	

Tubular sections					
Class	Section in bending and/or compression				
1	$d/t \leq 50\varepsilon^2$				
2	$d/t \leq 70\varepsilon^2$				
3	$d/t \leq 90\varepsilon^2$				
	NOTE For $d/t < 90\varepsilon^2$ see EN 1993-1-6				
$\varepsilon = \sqrt{235/f_y}$	f_y	235	275	355	420	460
	ε	1.00	0.92	0.81	0.75	0.71
	ε^2	1.00	0.85	0.66	0.56	0.51

Clause 6.2.2.4

1. The overall classification is taken as the highest (least favourable) class of its component parts, with the exceptions that (i) cross-sections with Class 3 webs and Class 1 or 2 flanges may be classified as Class 2 cross-sections with an effective web (defined in ***clause 6.2.2.4***) and (ii) in cases where the web is assumed to carry shear force only (and not to contribute to the bending or axial resistance of the cross-section), the classification may be based on that of the flanges (but Class 1 is not allowed).
2. The overall classification is defined by quoting both the flange and the web classification.

Figure 5.7. Definition of compression width c for common cases. (a) Outstand flanges. (b) Internal compression parts

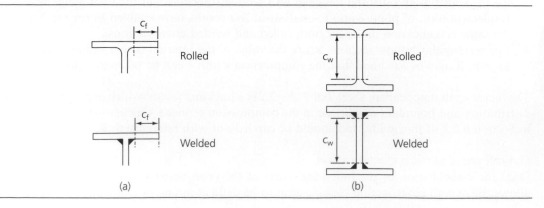

(a)   (b)

Class 4 cross-sections

Class 4 cross-sections (see *clause 6.2.2.5*) contain slender elements that are susceptible to local buckling in the elastic material range. Allowance for the reduction in resistance of Class 4 cross-sections as a result of local buckling is made by assigning effective widths to the Class 4 compression elements. The formulae for calculating effective widths are not contained in Part 1.1 of Eurocode 3; instead, the designer is directed to Part 1.3 for cold-formed sections, to Part 1.5 for hot-rolled and fabricated sections, and to Part 1.6 for circular hollow sections. The calculation of effective properties for Class 4 cross-sections is described in detail in Section 6.2.2 of this guide.

Clause 6.2.2.5

Classification under combined bending and axial force

Cross-sections subjected to combined bending and compression should be classified based on the actual stress distribution of the combined loadings. For simplicity, an initial check may be carried under the most severe loading condition of pure axial compression; if the resulting section classification is either Class 1 or Class 2, nothing is to be gained by conducting additional calculations with the actual pattern of stresses. However, if the resulting section classification is Class 3 or 4, it is advisable for economy to conduct a more precise classification under the combined loading.

For checking against the Class 1 and 2 cross-section slenderness limits, a plastic distribution of stress may be assumed, whereas an elastic distribution may be assumed for the Class 3 limits. To apply the classification limits from *Table 5.2* for a cross-section under combined bending and compression first requires the calculation of α (for Class 1 and 2 limits) and ψ (for Class 3 limits), where α is the ratio of the compressed width to the total width of an element and ψ is the ratio of end stresses (Figure 5.8). The topic of section classification under combined loading is covered in detail elsewhere (Davison and Owens, 2011). For the common case of an I or H section subjected to compression and major axis bending, where the neutral axis lies within the web, α, the ratio of the compressed width to the total width of the element, can be calculated using the equation

$$\alpha = \frac{1}{c_w}\left(\frac{h}{2} + \frac{1}{2}\frac{N_{Ed}}{t_w f_y} - (t_f + r)\right) \leq 1 \qquad (D5.4)$$

where c_w is the compressed width of the web (see Figure 5.8) and N_{Ed} is the axial compression force; use of the plastic stress distribution also requires that the compression flange is at least Class 2. The ratio of end stresses ψ (required for checking against the Class 3 limits) may be determined by superimposing the elastic bending stress distribution with the uniform compression stress distribution.

Design rules for verifying the resistance of structural components under combined bending and axial compression are given in *clause 6.2.9* for cross-sections and *clause 6.3.3* for members. An example demonstrating cross-section classification for a section under combined bending and compression is given below.

Clause 6.2.9
Clause 6.3.3

Figure 5.8. Definitions of α and ψ for classification of cross-sections under combined bending and compression. (a) Class 1 and 2 cross-sections. (b) Class 3 cross-sections

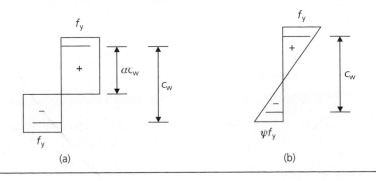

Example 5.1: cross-section classification under combined bending and compression

A member is to be designed to carry combined bending and axial load. In the presence of a major axis (y–y) bending moment and an axial force of 300 kN, determine the cross-section classification of a 406 × 178 × 54 UKB in grade S275 steel (Figure 5.9).

Figure 5.9. Section properties for 406 × 178 × 54 UKB

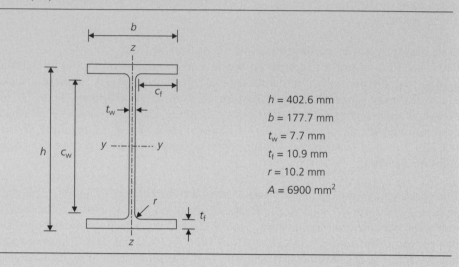

$h = 402.6$ mm
$b = 177.7$ mm
$t_w = 7.7$ mm
$t_f = 10.9$ mm
$r = 10.2$ mm
$A = 6900$ mm^2

For a nominal material thickness ($t_f = 10.9$ mm and $t_w = 7.7$ mm) of less than or equal to 16 mm the nominal value of yield strength f_y for grade S275 steel is found from EN 10025-2 to be 275 N/mm^2.

From *clause 3.2.6*: $E = 210\,000$ N/mm^2

Section properties

First, classify the cross-section under the most severe loading condition of pure compression to determine whether anything is to be gained by more precise calculations.

Clause 5.5.2

Cross-section classification under pure compression (clause 5.5.2)

$$\varepsilon = \sqrt{235/f_y} = \sqrt{235/275} = 0.92$$

Outstand flanges (*Table 5.2*, sheet 2):

$c_f = (b - t_w - 2r)/2 = 74.8$ mm

$c_f/t_f = 74.8/10.9 = 6.86$

Limit for Class 1 flange $= 9\varepsilon = 8.32$

$8.32 > 6.86$ ∴ flange is Class 1

Web – internal part in compression (*Table 5.2*, sheet 1):

$c_w = h - 2t_f - 2r = 360.4$ mm

$c_w/t_w = 360.4/7.7 = 46.81$

Limit for Class 3 web $= 42\varepsilon = 38.8$

$38.8 < 46.81$ ∴ web is Class 4

Under pure compression, the overall cross-section classification is therefore Class 4. Calculation and material efficiency are therefore to be gained by using a more precise approach.

Clause 5.5.2

Cross-section classification under combined loading (clause 5.5.2)
Flange classification remains as Class 1.

Web – internal part in bending and compression (*Table 5.2, sheet 1*).

From *Table 5.2* (sheet 1), for a Class 2 cross-section:

when $\alpha > 0.5$:

$$\frac{c_w}{t_w} \leq \frac{456\varepsilon}{13\alpha - 1}$$

when $\alpha \leq 0.5$:

$$\frac{c_w}{t_w} \leq \frac{41.5\varepsilon}{\alpha}$$

where α may be determined from equation (D5.4), for an I or H section where the neutral axis lies within the web.

$$\alpha = \frac{1}{c_w}\left(\frac{h}{2} + \frac{1}{2}\frac{N_{Ed}}{t_w f_y} - (t_f + r)\right) \leq 1$$

$$= \frac{1}{360.4}\left(\frac{402.6}{2} + \frac{1}{2}\frac{300\,000}{7.7 \times 275} - (10.9 + 10.2)\right) \qquad \text{(D5.4)}$$

$$= 0.70$$

$\therefore$ limit for a Class 2 web $= \dfrac{456\varepsilon}{13\alpha - 1} = 52.33$

$52.33 > 46.81 \qquad \therefore$ web is Class 2

Overall cross-section classification under the combined loading is therefore Class 2.

Conclusion
For this cross-section, a maximum axial force of 411 kN may be sustained in combination with a major axis bending moment, whilst remaining within the limits of a Class 2 section.

Cross-section and member resistance to combined bending and axial force is covered in Sections 6.2.9 and 6.3.3 of this guide, respectively.

5.6.   Cross-section requirements for plastic global analysis

For structures designed on the basis of a plastic global analysis, a series of requirements is placed upon the cross-sections of the constituent members, to ensure that the structural behaviour accords with the assumptions of the analysis. For cross-sections, in essence, this requires the provision of adequate rotation capacity at the plastic hinges.

Clause 5.6 deems that, for a uniform member, a cross-section has sufficient rotation capacity provided both of the following requirements are satisfied:

Clause 5.6

1. the member has a Class 1 cross-section at the plastic hinge location
2. web stiffeners are provided within a distance along the member of $h/2$ from the plastic hinge location, in cases where a transverse force that exceeds 10% of the shear resistance of the cross-section is applied at the plastic hinge location.

Additional criteria are specified in *clause 5.6(3)* for non-uniform members, where the cross-section varies along the length. Allowance for fastener holes in tension should be made with reference to *clause 5.6(4)*. Guidance on member requirements for plastically designed structures is given in Chapter 11 of this guide.

Clause 5.6(3)

Clause 5.6(4)

REFERENCES

Davison B and Owens GW (2011) *The Steel Designers' Manual*, 7th ed. Steel Construction Institute, Ascot, and Blackwell, Oxford.

SCI (2005) NCCI SN004: Calculation of alpha-cr. http://www.steel-ncci.co.uk

Designers' Guide to Eurocode 3: Design of Steel Buildings, 2nd ed.
ISBN 978-0-7277-4172-1

ICE Publishing: All rights reserved
doi: 10.1680/dsb.41721.031

Chapter 6
Ultimate limit states

This chapter concerns the subject of cross-section and member design at ultimate limit states. The material in this chapter is covered in *Section 6* of EN 1993-1-1 and the following clauses are addressed:

- General   *Clause 6.1*
- Resistance of cross-sections   *Clause 6.2*
- Buckling resistance of members   *Clause 6.3*
- Uniform built-up compression members   *Clause 6.4*

Unlike BS 5950: Part 1, which is largely self-contained, EN 1993-1-1 is not a stand-alone document. This is exemplified in *Section 6*, where reference is frequently made to other parts of the code – for example, the determination of effective widths for Class 4 cross-sections is not covered in Part 1.1, instead the designer should refer to Part 1.5, *Plated Structural Elements*. Although Eurocode 3 has come under some criticism for this approach, the resulting Part 1.1 is slimline while catering for the majority of structural steel design situations.

6.1.   General
In the structural Eurocodes, partial factors γ_{Mi} are applied to different components in various situations to reduce their resistances from characteristic values to design values (or, in practice, to ensure that the required level of safety is achieved). The uncertainties (material, geometry, modelling, etc.) associated with the prediction of resistance for a given case, as well as the chosen resistance model, dictate the value of γ_M that is to be applied. Partial factors are discussed in Section 2.4 of this guide, and in more detail in EN 1990 and elsewhere.[2] γ_{Mi} factors assigned to particular resistances in EN 1993-1-1 are as follows:

- resistance of cross-sections, γ_{M0}
- resistance of members to buckling (assessed by checks in *clause 6.3*), γ_{M1} *Clause 6.3*
- resistance of cross-sections in tension to fracture, γ_{M2}.

Numerical values for the partial factors recommended by Eurocode 3 for buildings are given in Table 6.1. However, for buildings to be constructed in the UK, values from *clause NA.2.15* of the   *Clause NA.2.15*
UK National Annex should be applied; these are also given in Table 6.1.

Clause 6.2
Clauses 6.2 and *6.3* cover the resistance of cross-sections and the resistance of members, respec-   *Clause 6.3*
tively. In general, both cross-sectional and member checks must be performed.

6.2.   Resistance of cross-sections
6.2.1   General
Prior to determining the resistance of a cross-section, the cross-section should be classified in accordance with *clause 5.5*. Cross-section classification is described in detail in Section 5.5 of   *Clause 5.5*
this guide. *Clause 6.2* covers the resistance of cross-sections including the resistance to tensile   *Clause 6.2*
fracture at net sections (where holes for fasteners exist).

Clause 6.2.1(4) allows the resistance of all cross-sections to be verified elastically (provided   *Clause 6.2.1(4)*
effective properties are used for Class 4 sections). For this purpose, the familiar von Mises
yield criterion is offered in *clause 6.2.1(5)*, as given by *equation (6.1)*, whereby the interaction   *Clause 6.2.1(5)*
of the local stresses should not exceed the yield stress (divided by the partial factor γ_{M0}) at

31

Table 6.1. Numerical values of partial factors γ_M for buildings

Partial factor γ_M	Eurocode 3	UK National Annex
γ_{M0}	1.00	1.00
γ_{M1}	1.00	1.00
γ_{M2}	1.25	1.10

any critical point:

$$\left(\frac{\sigma_{x,Ed}}{f_y/\gamma_{M0}}\right)^2 + \left(\frac{\sigma_{z,Ed}}{f_y/\gamma_{M0}}\right)^2 - \left(\frac{\sigma_{x,Ed}}{f_y/\gamma_{M0}}\right)\left(\frac{\sigma_{z,Ed}}{f_y/\gamma_{M0}}\right) + 3\left(\frac{\tau_{Ed}}{f_y/\gamma_{M0}}\right)^2 \leq 1 \qquad (6.1)$$

where

$\sigma_{x,Ed}$ is the design value of the local longitudinal stress at the point of consideration

$\sigma_{z,Ed}$ is the design value of the local transverse stress at the point of consideration

τ_{Ed} is the design value of the local shear stress at the point of consideration.

Although *equation (6.1)* is provided, the majority of design cases can be more efficiently and effectively dealt with using the interaction expressions given throughout *Section 6* of the code, since these are based on the readily available member forces and moments, and they allow more favourable (partially plastic) interactions.

6.2.2   Section properties

General

Clause 6.2.2

Clause 6.2.2 covers the calculation of cross-sectional properties. Provisions are made for the determination of gross and net areas, effective properties for sections susceptible to shear lag and local buckling (Class 4 elements), and effective properties for the special case where cross-sections with Class 3 webs and Class 1 or 2 flanges are classified as (effective) Class 2 cross-sections.

Gross and net areas

The gross area of a cross-section is defined in the usual way and utilises nominal dimensions. No reduction to the gross area is made for fastener holes, but allowance should be made for larger openings, such as those for services. Note that Eurocode 3 uses the generic term 'fasteners' to cover bolts, rivets and pins.

The method for calculating the net area of a cross-section in EN 1993-1-1 is essentially the same as that described in BS 5950: Part 1, with marginally different rules for sections such as angles with fastener holes in both legs. In general, the net area of the cross-section is taken as the gross area less appropriate deductions for fastener holes and other openings.

For a non-staggered arrangement of fasteners, for example as shown in Figure 6.1, the total area to be deducted should be taken as the sum of the sectional areas of the holes on any line (A–A) perpendicular to the member axis that passes through the centreline of the holes.

For a staggered arrangement of fasteners, for example as shown in Figure 6.2, the total area to be deducted should be taken as the greater of:

1. the maximum sum of the sectional areas of the holes on any line (A–A) perpendicular to the member axis

2. $t\left(nd_0 - \sum \dfrac{s^2}{4p}\right)$

measured on any diagonal or zig-zag line (A–B), where

- Rolled rectangular hollow section of uniform thickness, load parallel to the width:

$A_v = Ab/(b + h)$

- Circular hollow section and tubes of uniform thickness:

$A_v = 2A/\pi$

where

A is the cross-sectional area
b is the overall section breadth
h is the overall section depth
h_w is the overall web depth (measured between the flanges)
r is the root radius
t_f is the flange thickness
t_w is the web thickness (taken as the minimum value if the web is not of constant thickness)
η is a factor applied to the shear area, specified in *clause NA.2.4* of the UK National Annex to EN 1993-1-5 as equal to 1.00 for all steels.

Clause NA.2.4

The code also provides expressions in *clause 6.2.6(4)* for checking the elastic shear resistance of a cross-section, where the distribution of shear stresses is calculated assuming elastic material behaviour (see Figure 6.11). This check need only be applied to unusual sections that are not addressed in *clause 6.2.6(2)*, or in cases where plasticity is to be avoided, such as where repeated load reversal occurs.

Clause 6.2.6(4)

Clause 6.2.6(2)

The resistance of the web to shear buckling should also be checked, although this is unlikely to affect cross-sections of standard hot-rolled proportions. Shear buckling need not be considered provided:

$$\frac{h_w}{t_w} \leq 72\frac{\varepsilon}{\eta} \qquad \text{for unstiffened webs} \tag{D6.5}$$

$$\frac{h_w}{t_w} \leq 31\frac{\varepsilon}{\eta}\sqrt{k_\tau} \qquad \text{for webs with intermediate stiffeners} \tag{D6.6}$$

where

$$\varepsilon = \sqrt{\frac{235}{f_y}}$$

k_τ is a shear buckling coefficient defined in Annex A.3 of EN 1993-1-5.

For cross-sections that fail to meet the criteria of equations (D6.5) and (D6.6), reference should be made to clause 5.2 of EN 1993-1-5, to determine shear buckling resistance. Rules for combined shear force and torsion are provided in *clause 6.2.7(9)*.

Clause 6.2.7(9)

Example 6.4: shear resistance
Determine the shear resistance of a 229 × 89 rolled channel section in grade S275 steel loaded parallel to the web.

Section properties
The section properties are given in Figure 6.12.

For a nominal material thickness ($t_f = 13.3$ mm and $t_w = 8.6$ mm) of less than or equal to 16 mm the nominal value of yield strength f_y for grade S275 steel is found from EN 10025-2 to be 275 N/mm^2.

Figure 6.12. Section properties for a 229 × 89 mm rolled channel section

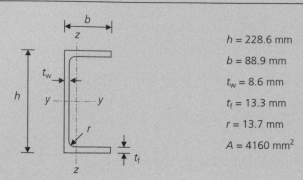

$h = 228.6$ mm

$b = 88.9$ mm

$t_w = 8.6$ mm

$t_f = 13.3$ mm

$r = 13.7$ mm

$A = 4160$ mm^2

Clause 6.2.6
Clause 6.2.6

Shear resistance (clause 6.2.6)

Shear resistance is determined according to *clause 6.2.6*:

$$V_{pl,Rd} = \frac{A_v(f_y/\sqrt{3})}{\gamma_{M0}} \qquad (6.18)$$

Clause NA.2.15

A numerical value of $\gamma_{M0} = 1.00$ is specified in *clause NA.2.15* of the UK National Annex.

Shear area A$_v$

For a rolled channel section, loaded parallel to the web, the shear area is given by

$$A_v = A - 2bt_f + (t_w + r)t_f$$

$$= 4160 - (2 \times 88.9 \times 13.3) + (8.6 + 13.7) \times 13.3$$

$$= 2092 \text{ mm}^2$$

$$\therefore V_{pl,Rd} = \frac{2092 \times (275/\sqrt{3})}{1.00} = 332\,000 \text{ N} = 332 \text{ kN}$$

Shear buckling

Shear buckling need not be considered, provided:

$$\frac{h_w}{t_w} \leq 72\frac{\varepsilon}{\eta} \qquad \text{for unstiffened webs}$$

$$\varepsilon = \sqrt{235/f_y} = \sqrt{235/275} = 0.92$$

Clause NA.2.4

$\eta = 1.0$ from *clause NA.2.4* of the UK National Annex to EN 1993-1-5.

$$72\frac{\varepsilon}{\eta} = 72 \times \frac{0.92}{1.0} = 66.6$$

Actual $h_w/t_w = (h - 2t_f)/t_w = [228.6 - (2 \times 13.3)]/8.6 = 23.5$

$$23.5 \leq 66.6 \qquad \therefore \text{ no shear buckling check required}$$

Conclusion

The shear resistance of a 229 × 89 rolled channel section in grade S275 steel loaded parallel to the web is 332 kN.

6.2.7   Torsion

The resistance of cross-sections to torsion is covered in *clause 6.2.7*. Torsional loading can arise in two ways: either due to an applied torque (pure twisting) or due to transverse load applied eccentrically to the shear centre of the cross-section (twisting plus bending). In engineering structures, it is the latter that is the most common, and pure twisting is relatively unusual. Consequently, *clauses 6.2.7*, *6.2.8* and *6.2.10* provide guidance for torsion acting in combination with other effects (bending, shear and axial force).

The torsional moment design effect T_{Ed} is made up of two components: the Saint Venant torsion $T_{t,Ed}$ and the warping torsion $T_{w,Ed}$.

Saint Venant torsion is the uniform torsion that exists when the rate of change of the angle of twist along the length of a member is constant. In such cases, the longitudinal warping deformations (that accompany twisting) are also constant, and the applied torque is resisted by a single set of shear stresses, distributed around the cross-section.

Warping torsion exists where the rate of change of the angle of twist along the length of a member is not constant; in which case, the member is said to be in a state of non-uniform torsion. Such non-uniform torsion may occur either as a result of non-uniform loading (i.e. varying torque along the length of the member) or due to the presence of longitudinal restraint to the warping deformations. For non-uniform torsion, longitudinal direct stresses and an additional set of shear stresses arise.

Therefore, as noted in *clause 6.2.7(4)*, there are three sets of stresses that should be considered:

- shear stresses $\tau_{t,Ed}$ due to the Saint Venant torsion
- shear stresses $\tau_{w,Ed}$ due to the warping torsion
- longitudinal direct stresses $\sigma_{w,Ed}$ due to the warping.

Depending on the cross-section classification, torsional resistance may be verified plastically with reference to *clause 6.2.7(6)*, or elastically by adopting the yield criterion of *equation (6.1)* (see *clause 6.2.1(5)*). Detailed guidance on the design of members subjected to torsion is available (Trahair *et al.*, 2008).

Clause 6.2.7(7) allows useful simplifications for the design of torsion members. For closed-section members (such as cylindrical and rectangular hollow sections), whose torsional rigidities are very large, Saint Venant torsion dominates, and warping torsion may be neglected. Conversely, for open sections, such as I- or H-sections, whose torsional rigidities are low, Saint Venant torsion may be neglected.

For the case of combined shear force and torsional moment, *clause 6.2.7(9)* defines a reduced plastic shear resistance $V_{pl,T,Rd}$, which must be demonstrated to be greater than the design shear force V_{Ed}.

$V_{pl,T,Rd}$ may be derived from *equations (6.26)* to *(6.28)*:

- for an I- or H-section

$$V_{pl,T,Rd} = \sqrt{1 - \frac{\tau_{t,Ed}}{1.25(f_y/\sqrt{3})/\gamma_{M0}}} V_{pl,Rd} \tag{6.26}$$

- for a channel section

$$V_{pl,T,Rd} = \left(\sqrt{1 - \frac{\tau_{t,Ed}}{1.25(f_y/\sqrt{3})/\gamma_{M0}}} - \frac{\tau_{w,Ed}}{(f_y/\sqrt{3})/\gamma_{M0}} \right) V_{pl,Rd} \tag{6.27}$$

- for a structural hollow section:

$$V_{pl,T,Rd} = \left(1 - \frac{\tau_{t,Ed}}{(f_y/\sqrt{3})/\gamma_{M0}} \right) V_{pl,Rd} \tag{6.28}$$

where $\tau_{t,Ed}$ and $\tau_{w,Ed}$ are defined above and $V_{pl,Rd}$ is obtained from *clause 6.2.6*.

Clause 6.2.7

Clause 6.2.7
Clause 6.2.8
Clause 6.2.10

Clause 6.2.7(4)

Clause 6.2.7(6)
Clause 6.2.1(5)

Clause 6.2.7(7)

Clause 6.2.7(9)

Clause 6.2.6

6.2.8 Bending and shear

Bending moments and shear forces acting in combination on structural members is commonplace. However, in the majority of cases (and particularly when standard rolled sections are adopted) the effect of shear force on the moment resistance is negligible and may be ignored – *clause 6.2.8(2)* states that provided the applied shear force is less than half the plastic shear resistance of the cross-section its effect on the moment resistance may be neglected. The exception to this is where shear buckling reduces the resistance of the cross-section, as described in Section 6.2.6 of this guide.

Clause 6.2.8(2)

For cases where the applied shear force is greater than half the plastic shear resistance of the cross-section, the moment resistance should be calculated using a reduced design strength for the shear area, given by *equation (6.29)*:

$$f_{yr} = (1 - \rho)f_y \tag{6.29}$$

where ρ is defined by equation (D6.7),

$$\rho = \left(\frac{2V_{Ed}}{V_{pl,Rd}} - 1\right)^2 \quad \text{(for } V_{Ed} > 0.5V_{pl,Rd}) \tag{D6.7}$$

$V_{pl,Rd}$ may be obtained from *clause 6.2.6*, and when torsion is present $V_{pl,Rd}$ should be replaced by $V_{pl,T,Rd}$, obtained from *clause 6.2.7*.

Clause 6.2.6
Clause 6.2.7

An alternative to the reduced design strength for the shear area, defined by *equation (6.29)*, which involves somewhat tedious calculations, is *equation (6.30)*. *Equation (6.30)* may be applied to the common situation of an I section (with equal flanges) subjected to bending about the major axis. In this case, the reduced design plastic resistance moment allowing for shear is given by

$$M_{y,V,Rd} = \frac{(W_{pl,y} - \rho A_w^2/4t_w)f_y}{\gamma_{M0}} \quad \text{but } M_{y,V,Rd} \leq M_{y,c,Rd} \tag{6.30}$$

Clause 6.2.5

where ρ is defined by equation (D6.7), $M_{y,c,Rd}$ may be obtained from *clause 6.2.5* and $A_w = h_w t_w$.

An example of the application of the cross-section rules for combined bending and shear force is given in Example 6.5.

Example 6.5: cross-section resistance under combined bending and shear

A short-span (1.4 m), simply supported, laterally restrained beam is to be designed to carry a central point load of 1050 kN, as shown in Figure 6.13.

Figure 6.13. General arrangement and loading

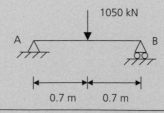

The arrangement of Figure 6.13 results in a maximum design shear force V_{Ed} of 525 kN and a maximum design bending moment M_{Ed} of 367.5 kN m.

In this example a $406 \times 178 \times 74$ UKB in grade S275 steel is assessed for its suitability for this application.

Section properties
The section properties are set out in Figure 6.14.

Figure 6.14. Section properties for a $406 \times 178 \times 74$ UKB

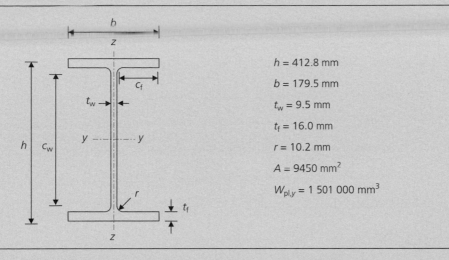

$h = 412.8$ mm

$b = 179.5$ mm

$t_w = 9.5$ mm

$t_f = 16.0$ mm

$r = 10.2$ mm

$A = 9450$ mm^2

$W_{pl,y} = 1\,501\,000$ mm^3

For a nominal material thickness ($t_f = 16.0$ mm and $t_w = 9.5$ mm) of less than or equal to 16 mm the nominal value of yield strength f_y for grade S275 steel is found from EN 10025-2 to be 275 N/mm^2.

From *clause 3.2.6*:

Clause 3.2.6

$$E = 210\,000 \text{ N/mm}^2$$

Cross-section classification (clause 5.5.2)

Clause 5.5.2

$$\varepsilon = \sqrt{235/f_y} = \sqrt{235/275} = 0.92$$

Outstand flange in compression (*Table 5.2*, sheet 2):

$$c_f = (b - t_w - 2r)/2 = 74.8 \text{ mm}$$

$$c_f/t_f = 74.8/16.0 = 4.68$$

Limit for Class 1 flange $= 9\varepsilon = 8.32$

$8.32 > 4.68$ $\therefore$ flange is Class 1

Web – internal part in bending (*Table 5.2*, sheet 1):

$$c_w = h - 2t_f - 2r = 360.4 \text{ mm}$$

$$c_w/t_w = 360.4/9.5 = 37.94$$

Limit for Class 1 web $= 72\varepsilon = 66.56$

$66.56 > 37.94$ $\therefore$ web is Class 1

Therefore, the overall cross-section classification is Class 1.

Bending resistance of cross-section (clause 6.2.5)

Clause 6.2.5

$$M_{c,y,Rd} = \frac{W_{pl,y}f_y}{\gamma_{M0}} \qquad \text{for Class 1 or 2 cross-sections} \qquad (6.13)$$

The design bending resistance of the cross-section

$$M_{c,y,Rd} = \frac{1501 \times 10^3 \times 275}{1.00} = 412 \times 10^6 \text{ N mm} = 412 \text{ kN m}$$

412 kN m > 367.5 kN m $\therefore$ cross-section resistance in bending is acceptable

Clause 6.2.6

***Shear resistance of cross-section** (clause 6.2.6)*

$$V_{\text{pl,Rd}} = \frac{A_v(f_y/\sqrt{3})}{\gamma_{\text{M0}}} \tag{6.18}$$

For a rolled I section, loaded parallel to the web, the shear area A_v is given by

$$A_v = A - 2bt_f + (t_w + 2r)t_f \text{ (but not less than } \eta h_w t_w)$$

Clause NA.2.4

$\eta = 1.0$ from ***clause NA.2.4*** of the UK National Annex to EN 1993-1-5.

$$h_w = (h - 2t_f) = 412.8 - (2 \times 16.0) = 380.8 \text{ mm}$$

$$\therefore A_v = 9450 - (2 \times 179.5 \times 16.0) + [9.5 + (2 \times 10.2)] \times 16.0$$

$$= 4184 \text{ mm}^2 \text{ (but not less than } 1.0 \times 380.8 \times 9.5 = 3618 \text{ mm}^2)$$

$$V_{\text{pl,Rd}} = \frac{4184 \times (275/\sqrt{3})}{1.00} = 664\,300 \text{ N} = 664.3 \text{ kN}$$

Shear buckling need not be considered, provided

$$\frac{h_w}{t_w} \leq 72\frac{\varepsilon}{\eta} \qquad \text{for unstiffened webs}$$

$$72\frac{\varepsilon}{\eta} = 72 \times \frac{0.92}{1.0} = 66.6$$

Actual $h_w/t_w = 380.8/9.5 = 40.1$

$$40.1 \leq 66.6 \qquad \therefore \text{ no shear buckling check required}$$

$$664.3 > 525 \text{ kN} \qquad \therefore \text{ shear resistance is acceptable}$$

Clause 6.2.8

***Resistance of cross-section to combined bending and shear** (clause 6.2.8)*
The applied shear force is greater than half the plastic shear resistance of the cross-section, therefore a reduced moment resistance $M_{y,V,Rd}$ must be calculated. For an I section (with equal flanges) and bending about the major axis, ***clause 6.2.8(5)*** and *equation (6.30)* may be utilised.

Clause 6.2.8(5)

$$M_{y,V,Rd} = \frac{(W_{\text{pl,y}} - \rho A_w^2/4t_w)f_y}{\gamma_{\text{M0}}} \qquad \text{but } M_{y,V,Rd} \leq M_{y,c,Rd} \tag{6.30}$$

$$\rho = \left(\frac{2V_{\text{Ed}}}{V_{\text{pl,Rd}}} - 1\right)^2 = \left(\frac{2 \times 525}{689.2} - 1\right)^2 = 0.27 \tag{D6.7}$$

$$A_w = h_w t_w = 380.8 \times 9.5 = 3617.6 \text{ mm}^2$$

$$\Rightarrow M_{y,V,Rd} = \frac{(1\,501\,000 - 0.27 \times 3617.6^2/4 \times 9.5) \times 275}{1.0} = 386.8 \text{ kNm}$$

$$386.8 \text{ kN m} > 367.5 \text{ kN m} \qquad \therefore \text{ cross-section resistance to combined bending}$$
$$\text{and shear is acceptable}$$

Conclusion
A $406 \times 178 \times 74$ UKB in grade S275 steel is suitable for the arrangement and loading shown by Figure 6.13.

6.2.9 Bending and axial force
The design of cross-sections subjected to combined bending and axial force is described in ***clause***

Clause 6.2.9

6.2.9. Bending may be about one or both principal axes, and the axial force may be tensile or compressive (with no difference in treatment). In dealing with the combined effects, Eurocode 3 prescribes different methods for designing Class 1 and 2, Class 3, and Class 4 cross-sections.

As an overview to the codified approach, for Class 1 and 2 sections, the basic principle is that the design moment should be less than the reduced moment capacity, reduced, that is, to take account of the axial load. For Class 3 sections, the maximum longitudinal stress due to the combined actions must be less than the yield stress, while for Class 4 sections the same criterion is applied but to a stress calculated based on effective cross-section properties.

As a conservative alternative to the methods set out in the following subsections, a simple linear interaction given below and in *equation (6.2)* may be applied to all cross-sections (*clause 6.2.1(7)*), although Class 4 cross-section resistances must be based on effective section properties (and any additional moments arising from the resulting shift in neutral axis should be allowed for). These additional moments necessitate the extended linear interaction expression given by *equation (6.44)* and discussed later.

Clause 6.2.1(7)

$$\frac{N_{Ed}}{N_{Rd}} + \frac{M_{y,Ed}}{M_{y,Rd}} + \frac{M_{z,Ed}}{M_{z,Rd}} \leq 1 \tag{6.2}$$

where N_{Rd}, $M_{y,Rd}$ and $M_{z,Rd}$ are the design cross-sectional resistances and should include any necessary reduction due to shear effects (*clause 6.2.8*). The intention of *equation (6.2)* is simply to allow a designer to generate a quick, approximate and safe solution, perhaps for the purposes of initial member sizing, with the opportunity to refine the calculations for final design.

Clause 6.2.8

Class 1 and 2 cross-sections: mono-axial bending and axial force
The design of Class 1 and 2 cross-sections subjected to mono-axial bending (i.e. bending about a single principal axis) and axial force is covered in *clause 6.2.9.1(5)*, while bi-axial bending (with or without axial forces) is covered in *clause 6.2.9.1(6)*.

Clause 6.2.9.1(5)
Clause 6.2.9.1(6)

In general, for Class 1 and 2 cross-sections (subjected to bending and axial forces), Eurocode 3 requires the calculation of a reduced plastic moment resistance $M_{N,Rd}$ to account for the presence of an applied axial force N_{Ed}. It should then be checked that the applied bending moment M_{Ed} is less than this reduced plastic moment resistance.

Clause 6.2.9.1(4) recognises that for small axial loads, the theoretical reduction in plastic moment capacity is essentially offset by material strain hardening, and may therefore be neglected. The clause states that for doubly symmetrical I- and H-sections, and other flanged sections subjected to axial force and major (y–y) axis bending moment, no reduction in the major axis plastic moment resistance is necessary provided both of the following criteria (*equations 6.33 and 6.34*) are met:

Clause 6.2.9.1(4)

$$N_{Ed} \leq 0.25 N_{pl,Rd} \tag{6.33}$$

$$N_{Ed} \leq \frac{0.5 h_w t_w f_y}{\gamma_{M0}} \tag{6.34}$$

And similarly, for doubly symmetrical I- and H-sections, rectangular rolled hollow sections and welded box sections subjected to axial force and minor (z–z) axis bending moment, no reduction in minor axis plastic moment resistance is necessary, provided

$$N_{Ed} \leq \frac{h_w t_w f_y}{\gamma_{M0}} \tag{6.35}$$

If the above criteria are not met, a reduced plastic moment resistance must be calculated using the expressions provided in *clause 6.2.9.1(5)* and given below.

Clause 6.2.9.1(5)

Reduced plastic moment resistance for:

1. Doubly-symmetrical I- and H-sections (hot-rolled or welded).

   Major (y–y) axis:

$$M_{N,y,Rd} = M_{pl,y,Rd} \frac{1-n}{1-0.5a} \qquad \text{but } M_{N,y,Rd} \leq M_{pl,y,Rd} \tag{6.36}$$

Minor (z–z) axis:

$$M_{\mathrm{N},z,\mathrm{Rd}} = M_{\mathrm{pl},z,\mathrm{Rd}} \qquad \text{for } n \leq a \tag{6.37}$$

$$M_{\mathrm{N},z,\mathrm{Rd}} = M_{\mathrm{pl},z,\mathrm{Rd}}\left[1 - \left(\frac{n-a}{1-a}\right)^2\right] \qquad \text{for } n > a \tag{6.38}$$

where

$$n = \frac{N_{\mathrm{Ed}}}{N_{\mathrm{pl},\mathrm{Rd}}}$$

is the ratio of applied load to plastic compression resistance of section and

$$a = \frac{A - 2bt_{\mathrm{f}}}{A} \qquad \text{but } a \leq 0.5$$

is the ratio of the area of the web to the total area.

2. Rectangular hollow sections of uniform thickness and welded box sections (with equal flanges and equal webs).

Major (y–y) axis:

$$M_{\mathrm{N},y,\mathrm{Rd}} = M_{\mathrm{pl},y,\mathrm{Rd}}\frac{1-n}{1-0.5a_{\mathrm{w}}} \qquad \text{but } M_{\mathrm{N},y,\mathrm{Rd}} \leq M_{\mathrm{pl},y,\mathrm{Rd}} \tag{6.39}$$

Minor (z–z) axis:

$$M_{\mathrm{N},z,\mathrm{Rd}} = M_{\mathrm{pl},z,\mathrm{Rd}}\frac{1-n}{1-0.5a_{\mathrm{f}}} \qquad \text{but } M_{\mathrm{N},z,\mathrm{Rd}} \leq M_{\mathrm{pl},z,\mathrm{Rd}} \tag{6.40}$$

where

$$a_{\mathrm{w}} = \frac{A - 2bt}{A} \qquad \text{but } a_{\mathrm{w}} \leq 0.5$$

for hollow sections

$$a_{\mathrm{w}} = \frac{A - 2bt_{\mathrm{f}}}{A} \qquad \text{but } a_{\mathrm{w}} \leq 0.5$$

for welded box sections

$$a_{\mathrm{f}} = \frac{A - 2ht}{A} \qquad \text{but } a_{\mathrm{f}} \leq 0.5$$

for hollow sections

$$a_{\mathrm{f}} = \frac{A - 2ht_{\mathrm{w}}}{A} \qquad \text{but } a_{\mathrm{f}} \leq 0.5$$

or welded box sections.

Example 6.6: cross-section resistance under combined bending and compression

A member is to be designed to carry a combined major axis bending moment and an axial force. In this example, a cross-sectional check is performed to determine the maximum bending moment that can be carried by a $457 \times 191 \times 98$ UKB in grade S275 steel, in the presence of an axial force of 1400 kN.

Section properties

The section properties are given in Figure 6.15.

Figure 6.15. Section properties for a 457 × 191 × 98 UKB

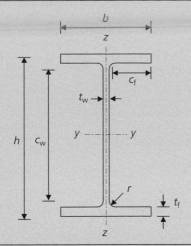

$h = 467.2$ mm

$b = 192.8$ mm

$t_w = 11.4$ mm

$t_f = 19.6$ mm

$r = 10.2$ mm

$A = 12\ 500$ mm^2

$W_{pl,y} = 2\ 230\ 000$ mm^3

For a nominal material thickness ($t_f = 19.6$ mm and $t_w = 11.4$ mm) of between 16 mm and 40 mm the nominal value of yield strength f_y for grade S275 steel is found from EN 10025-2 to be 265 N/mm^2.

From *clause 3.2.6*:

$E = 210\ 000$ N/mm^2

As in Example 5.1, first classify the cross-section under the most severe loading condition of pure compression to determine whether anything is to be gained by more precise calculations.

Cross-section classification under pure compression (clause 5.5.2)

$\varepsilon = \sqrt{235/f_y} = \sqrt{235/265} = 0.94$

Outstand flanges (*Table 5.2*, sheet 2):

$c_f = (b - t_w - 2r)/2 = 80.5$ mm

$c_f/t_f = 80.5/19.6 = 4.11$

Limit for Class 1 flange $= 9\varepsilon = 8.48$

$8.48 > 4.11$ ∴ flange is Class 1

Web – internal part in compression (*Table 5.2*, sheet 1):

$c_w = h - 2t_f - 2r = 407.6$ mm

$c_w/t_w = 407.6/11.4 = 35.75$

Limit for Class 2 web $= 38\varepsilon = 35.78$

$35.78 > 35.75$ ∴ web is Class 2

Under pure compression, the overall cross-section classification is therefore Class 2. Consequently, unlike Example 5.1, nothing is to be gained by using the more complex approach of considering the actual stress distribution.

Bending and axial force (clause 6.2.9.1)

No reduction to the plastic resistance moment due to the effect of axial force is required when both of the following criteria are satisfied:

$N_{Ed} \leq 0.25 N_{pl,Rd}$ (6.33)

Clause 3.2.6

Clause 5.5.2

Clause 6.2.9.1

and

$$N_{Ed} \leq \frac{0.5h_w t_w f_y}{\gamma_{M0}} \tag{6.34}$$

$$N_{Ed} = 1400 \text{ kN}$$

$$N_{pl,Rd} = \frac{A f_y}{\gamma_{M0}} = \frac{12\,500 \times 265}{1.0} = 3313 \text{ kN}$$

$$0.25 N_{pl,Rd} = 828.1 \text{ kN}$$

$$828.1 \text{ kN} < 1400 \text{ kN} \qquad \therefore \textit{ equation (6.33) is not satisfied}$$

$$\frac{0.5 h_w t_w f_y}{\gamma_{M0}} = \frac{0.5 \times [467.2 - (2 \times 19.6)] \times 11.4 \times 265}{1.0} = 646.5 \text{ kN}$$

$$646.5 \text{ kN} < 1400 \text{ kN} \qquad \therefore \textit{ equation (6.34) is not satisfied}$$

Therefore, allowance for the effect of axial force on the plastic moment resistance of the cross-section must be made.

Clause 6.2.9.1(5)

Reduced plastic moment resistance (clause 6.2.9.1(5))

$$M_{N,y,Rd} = M_{pl,y,Rd} \frac{1-n}{1-0.5a} \qquad \text{but } M_{N,y,Rd} \leq M_{pl,y,Rd} \tag{6.36}$$

where

$$n = N_{Ed}/N_{pl,Rd} = 1400/3313 = 0.42$$

$$a = (A - 2bt_f)/A = [12\,500 - (2 \times 192.8 \times 19.6)]/12\,500 = 0.40$$

$$M_{pl,y,Rd} = \frac{W_{pl} f_y}{\gamma_{M0}} = \frac{2\,230\,000 \times 265}{1.0} = 591.0 \text{ kN m}$$

$$\Rightarrow M_{N,y,Rd} = 591.0 \times \frac{1 - 0.42}{1 - (0.5 \times 0.40)} = 425.3 \text{ kN m}$$

Conclusion

Clause 6.2.9

In order to satisfy the cross-sectional checks of **clause 6.2.9**, the maximum bending moment that can be carried by a $457 \times 191 \times 98$ UKB in grade S275 steel, in the presence of an axial force of 1400 kN is 425.3 kN m.

Class 1 and 2 cross-sections: bi-axial bending with or without axial force

As in BS 5950: Part 1, EN 1993-1-1 treats bi-axial bending as a subset of the rules for combined bending and axial force. Checks for Class 1 and 2 cross-sections subjected to bi-axial bending, with or without axial forces, are set out in **clause 6.2.9.1(6)**. Although the simple linear inter-action expression of *equation (6.2)* may be used, *equation (6.41)* represents a more sophisticated convex interaction expression, which can result in significant improvements in efficiency:

Clause 6.2.9.1(6)

$$\left(\frac{M_{y,Ed}}{M_{N,y,Rd}}\right)^{\alpha} + \left(\frac{M_{z,Ed}}{M_{N,z,Rd}}\right)^{\beta} \leq 1 \tag{6.41}$$

Clause 6.2.9(6)

in which α and β are constants, as defined below. **Clause 6.2.9(6)** allows α and β to be taken as unity, thus reverting to a conservative linear interaction.

For I- and H-sections:

$$\alpha = 2 \qquad \text{and} \qquad \beta = 5n \qquad \text{but} \qquad \beta \geq 1$$

For circular hollow sections:

$$\alpha = 2 \qquad \text{and} \qquad \beta = 2$$

Figure 6.16. Bi-axial bending interaction curves

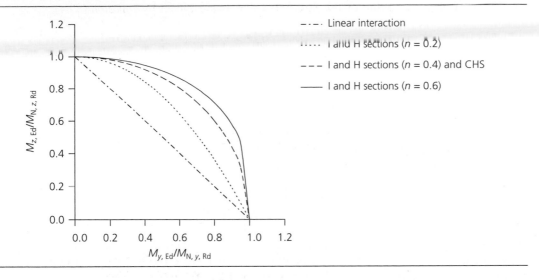

For rectangular hollow sections:

$$\alpha = \beta = \frac{1.66}{1 - 1.13n^2} \qquad \text{but } \alpha = \beta \leq 6$$

Figure 6.16 shows the bi-axial bending interaction curves (for Class 1 and 2 cross-sections) for some common cases.

Class 3 cross-sections: general

For Class 3 cross-sections, *clause 6.2.9.2* permits only a linear interaction of stresses arising from combined bending moments and axial force, and limits the maximum fibre stress (in the longitudinal x-direction of the member) to the yield stress, f_y divided by the partial factor γ_{M0}, as below:

$$\sigma_{x,\text{Ed}} = \frac{f_y}{\gamma_{M0}} \qquad (6.42)$$

Clause 6.2.9.2

As when considering compression and bending in isolation, allowances for fastener holes should be made in the unusual cases of slotted or oversized holes or where there are holes that contain no fasteners.

Class 4 cross-sections: general

As for Class 3 cross-sections, Class 4 sections subjected to combined bending and axial force (*clause 6.2.9.3*) are also designed based on a linear interaction of stresses, with the maximum fibre stress (in the longitudinal x-direction of the member) limited to the yield stress f_y divided by the partial factor γ_{M0}, as given by *equation (6.42)*.

Clause 6.2.9.3

However, for Class 4 cross-sections the stresses must be calculated on the basis of the effective properties of the section, and allowance must be made for the additional stresses resulting from the shift in neutral axis between the gross cross-section and the effective cross-section (see Figure 6.17, *clause 6.2.2.5(4)* and Chapter 13 of this guide).

Clause 6.2.2.5(4)

Figure 6.17. Shift in neutral axis from (a) gross to (b) effective cross-section

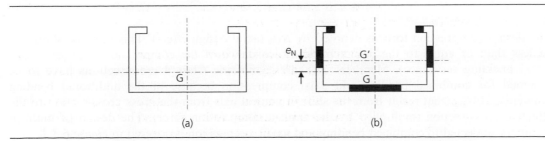

The resulting interaction expression that satisfies *equation (6.42)*, and includes the bending moments induced as a result of the shift in neutral axis, is given by *equation (6.44)*:

$$\frac{N_{\mathrm{Ed}}}{A_{\mathrm{eff}}\,f_{\mathrm{y}}/\gamma_{\mathrm{M0}}} + \frac{M_{\mathrm{y,Ed}} + N_{\mathrm{Ed}}e_{\mathrm{N}y}}{W_{\mathrm{eff},y,\min}\,f_{\mathrm{y}}/\gamma_{\mathrm{M0}}} + \frac{M_{\mathrm{z,Ed}} + N_{\mathrm{Ed}}e_{\mathrm{N}z}}{W_{\mathrm{eff},z,\min}\,f_{\mathrm{y}}/\gamma_{\mathrm{M0}}} \leq 1 \qquad (6.44)$$

where

A_{eff} is the effective area of the cross-section under pure compression

$W_{\mathrm{eff,min}}$ is the effective section modulus about the relevant axis, based on the extreme fibre that reaches yield first

e_{N} is the shift in the relevant neutral axis.

6.2.10. Bending, shear and axial force

Clause 6.2.10
Clause 6.2.6

Clause 6.2.9

The design of cross-sections subjected to combined bending, shear and axial force is covered in *clause 6.2.10*. However, provided the shear force V_{Ed} is less than 50% of the design plastic shear resistance $V_{\mathrm{pl,Rd}}$ (*clause 6.2.6*), and provided that shear buckling is not a concern (see Section 6.2.8 of this guide and clause 5.1 of EN 1993-1-5), then the cross-section need only satisfy the requirements for bending and axial force (*clause 6.2.9*).

Clause 6.2.8
Clause 6.2.9

In cases where the shear force does exceed 50% of the design plastic shear resistance of the cross-section, then a reduced yield strength should be calculated for the shear area (as described in *clause 6.2.8*); the cross-section, with a reduced strength shear area, may subsequently be checked for bending and axial force according to *clause 6.2.9*. As an alternative to reducing the strength of the shear area, an equivalent reduction to the thickness is also allowed; this may simplify calculations.

6.3. Buckling resistance of members

Clause 6.3

Clause 6.3.1
Clause 6.3.2
Clause 6.3.3

Clause 6.3.1
Clause 6.3.2
Clause 6.3.3

Clause 6.3 covers the buckling resistance of members. Guidance is provided for uniform compression members susceptible to flexural, torsional and torsional–flexural buckling (see *clause 6.3.1*), uniform bending members susceptible to lateral torsional buckling (see *clause 6.3.2*), and uniform members subjected to a combination of bending and axial compression (see *clause 6.3.3*). For member design, no account need be taken for fastener holes at the member ends.

Clauses 6.3.1 to *6.3.3* are applicable to uniform members, defined as those with a constant cross-section along the full length of the member (and, additionally, in the case of compression members, the load should be applied uniformly). For non-uniform members, such as those with tapered sections, or for members with a non-uniform distribution of compression force along their length (which may arise, for example, where framing-in members apply forces but offer no significant lateral restraint), Eurocode 3 provides no design expressions for calculating buckling resistances; it is, however, noted that a second-order analysis using the member imperfections according to *clause 5.3.4* may be used to directly determine member buckling resistances.

Clause 5.3.4

6.3.1 Uniform members in compression

General

The Eurocode 3 approach to determining the buckling resistance of compression members is based on the same principles as that of BS 5950. Although minor technical differences exist, the primary difference between the two codes is in the presentation of the method.

Buckling resistance

The design compression force is denoted by N_{Ed} (axial design effect). This must be shown to be less than or equal to the design buckling resistance of the compression member, $N_{\mathrm{b,Rd}}$ (axial buckling resistance). Members with non-symmetric Class 4 cross-sections have to be designed for combined bending and axial compression because of the additional bending moments, ΔM_{Ed}, that result from the shift in neutral axis from the gross cross-section to the effective cross-section (multiplied by the applied compression force). The design of uniform members subjected to combined bending and axial compression is covered in *clause 6.3.3*.

Clause 6.3.3

Compression members with Class 1, 2 and 3 cross-sections and symmetrical Class 4 cross-sections follow the provisions of *clause 6.3.1*, and the design buckling resistance should be taken as

Clause 6.3.1

$$N_{b,Rd} = \frac{\chi A f_y}{\gamma_{M1}} \qquad \text{for Class 1, 2 and 3 cross-sections} \qquad (6.47)$$

$$N_{b,Rd} = \frac{\chi A_{eff} f_y}{\gamma_{M1}} \qquad \text{for (symmetric) Class 4 cross-sections} \qquad (6.48)$$

where χ is the reduction factor for the relevant buckling mode (flexural, torsional or torsional flexural). These buckling modes are discussed later in this section.

Buckling curves

The buckling curves defined by EN 1993-1-1 are equivalent to those set out in BS 5950: Part 1 in tabular form in Table 24 (with the exception of buckling curve a_0, which does not appear in BS 5950). Regardless of the mode of buckling, the basic formulations for the buckling curves remain unchanged, and are as given below:

$$\chi = \frac{1}{\Phi + \sqrt{\Phi^2 - \bar{\lambda}^2}} \qquad \text{but } \chi \le 1.0 \qquad (6.49)$$

where

$$\Phi = 0.5[1 + \alpha(\bar{\lambda} - 0.2) + \bar{\lambda}^2]$$

$$\bar{\lambda} = \sqrt{\frac{A f_y}{N_{cr}}} \qquad \text{for Class 1, 2 and 3 cross-sections}$$

$$\bar{\lambda} = \sqrt{\frac{A_{eff} f_y}{N_{cr}}} \qquad \text{for Class 4 cross-sections}$$

α is an imperfection factor
N_{cr} is the elastic critical buckling force for the relevant buckling mode based on the gross properties of the cross-section.

The non-dimensional slenderness $\bar{\lambda}$, as defined above, is in a generalised format requiring the calculation of the elastic critical force N_{cr} for the relevant buckling mode. The relevant buckling mode that governs design will be that with the lowest critical buckling force N_{cr}. Calculation of N_{cr}, and hence $\bar{\lambda}$, for the various buckling modes is described in the following section.

As shown in Figure 6.18, EN 1993-1-1 defines five buckling curves, labelled a_0, a, b, c and d. The shapes of these buckling curves are altered through the imperfection factor α; the five values of the imperfection factor α for each of these curves are given in *Table 6.1* of the code (reproduced here as Table 6.4). It is worth noting that as an alternative to using the buckling curve formulations described above, *clause 6.3.1.2(3)* allows the buckling reduction factor to be determined graphically directly from *Figure 6.4* of the code (reproduced here as Figure 6.18).

Clause 6.3.1.2(3)

From the shape of the buckling curves given in Figure 6.18 it can be seen, in all cases, that for values of non-dimensional slenderness $\bar{\lambda} \le 0.2$ the buckling reduction factor is equal to unity. This means that for compression members of stocky proportions ($\bar{\lambda} \le 0.2$, or, in terms of elastic critical forces, for $N_{Ed}/N_{cr} \le 0.04$) there is no reduction to the basic cross-section resistance. In this case, buckling effects may be ignored and only cross-sectional checks (*clause 6.2*) need be applied.

Clause 6.2

The choice as to which buckling curve (imperfection factor) to adopt is dependent upon the geometry and material properties of the cross-section and upon the axis of buckling. The appropriate buckling curve should be determined from Table 6.5 (*Table 6.2* of EN 1993-1-1), which is equivalent to the 'allocation of strut curve' table (Table 23) of BS 5950: Part 1.

Figure 6.18. EN 1993-1-1 buckling curves

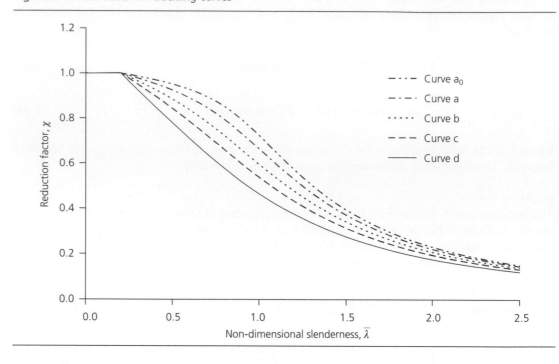

Non-dimensional slenderness for various buckling modes

Clause 6.3.1.3
Clause 6.3.1.4

EN 1993-1-1 provides guidance for flexural (*clause 6.3.1.3*), torsional (*clause 6.3.1.4*) and flexural–torsional (*clause 6.3.1.4*) buckling modes. For standard hot-rolled and welded structural cross-sections, flexural buckling is the predominant buckling mode, and hence governs design in the vast majority of cases.

Buckling modes with torsional components are generally limited to cold-formed members for two principal reasons:

- cold-formed cross-sections contain relatively thin material, and torsional stiffness is associated with the material thickness cubed
- the cold-forming process gives a predominance of open sections because these can be easily produced from flat sheet. Open sections have inherently low torsional stiffness.

Clause 6.3.1.3

Flexural buckling of a compression member is characterised by excessive lateral deflections in the plane of the weaker principal axis of the member. As the slenderness of the column increases, the load at which failure occurs reduces. Calculation of the non-dimensional slenderness for flexural buckling is covered in *clause 6.3.1.3*.

The non-dimensional slenderness $\bar{\lambda}$ is given by

$$\bar{\lambda} = \sqrt{\frac{A f_y}{N_{cr}}} = \frac{L_{cr}}{i} \frac{1}{\lambda_1} \qquad \text{for Class 1, 2 and 3 cross-sections} \qquad (6.50)$$

$$\bar{\lambda} = \sqrt{\frac{A_{eff} f_y}{N_{cr}}} = \frac{L_{cr}}{i} \frac{\sqrt{A_{eff}/A}}{\lambda_1} \qquad \text{for Class 4 cross-sections} \qquad (6.51)$$

Table 6.4. Imperfection factors for buckling curves (*Table 6.1* of EN 1993-1-1)

Buckling curve	a_0	a	b	c	d
Imperfection factor α	0.13	0.21	0.34	0.49	0.76

Table 6.5. Selection of buckling curve for a cross-section (*Table 6.2* of EN 1993-1-1)

Cross-section		Limits		Buckling about axis	Buckling curve S 235 S 275 S 355 S 420	Buckling curve S 460
Rolled sections		$h/b > 1.2$	$t_f \leq 40\,mm$	y–y z–z	a b	a_0 a_0
			$40\,mm < t_f \leq 100$	y–y z–z	b c	a a
		$h/b \leq 1.2$	$t_f \leq 100\,mm$	y–y z–z	b c	a a
			$t_f > 100\,mm$	y–y z–z	d d	c c
Welded I-sections		$t_f \leq 40\,mm$		y–y z–z	b c	b c
		$t_f > 40\,mm$		y–y z–z	c d	c d
Hollow sections		hot finished		any	a	a_0
		cold formed		any	c	c
Welded box sections		generally (except as below)		any	b	b
		thick welds: $a > 0.5t_f$ $b/t_f < 30$ $h/t_w < 30$		any	c	c
U-, T- and solid sections				any	c	c
L-sections				any	b	b

where

L_{cr} is the buckling length of the compression member in the plane under consideration, and is equivalent to the effective length L_E in BS 5950 (buckling lengths are discussed in the next section)

i is the radius of gyration about the relevant axis, determined using the gross properties of the cross-section (assigned the symbols r_x and r_y in BS 5950 for the radius of gyration about the major and minor axes, respectively)

$$\lambda_1 = \pi\sqrt{\frac{E}{f_y}} = 93.9\varepsilon \quad \text{and} \quad \varepsilon = \sqrt{\frac{235}{f_y}} \quad (f_y \text{ in N/mm}^2)$$

Clearly, the BS 5950 definition of slenderness ($\lambda = L_E/r_y$) is already 'non-dimensional', but the advantage of the Eurocode 3 definition of 'non-dimensional slenderness' $\bar{\lambda}$, which includes the material properties of the compression member through λ_1, is that all variables affecting the theoretical buckling load of a perfect pin-ended column are now present. This allows a more direct comparison of susceptibility to flexural buckling to be made for columns with varying material strength. Further, $\bar{\lambda}$ is useful for relating the column slenderness to the theoretical point at which the squash load and the Euler critical buckling load coincide, which always occurs at the value of non-dimensional slenderness $\bar{\lambda}$ equal to 1.0.

As stated earlier, flexural buckling is by far the most common buckling mode for conventional hot-rolled structural members. However, particularly for thin-walled and open sections, the designer should also check for the possibility that the torsional or torsional–flexural buckling resistance of a member may be less than the flexural buckling resistance. Torsional and torsional–flexural buckling are discussed further in Section 13.7 of this guide.

Clause 6.3.1.4

Calculation of the non-dimensional slenderness $\bar{\lambda}_T$ for torsional and torsional–flexural buckling is covered in *clause 6.3.1.4*, and should be taken as

$$\bar{\lambda}_T = \sqrt{\frac{Af_y}{N_{cr}}} \qquad \text{for Class 1, 2 and 3 cross-sections} \qquad (6.52)$$

$$\bar{\lambda}_T = \sqrt{\frac{A_{eff}f_y}{N_{cr}}} \qquad \text{for Class 4 cross-sections} \qquad (6.53)$$

where

$N_{cr} = N_{cr,TF}$ but $N_{cr} \leq N_{cr,T}$

$N_{cr,TF}$ is the elastic critical torsional–flexural buckling force (see Section 13.7 of this guide)

$N_{cr,T}$ is the elastic critical torsional buckling force (see Section 13.7 of this guide).

The generic definition of $\bar{\lambda}_T$ is the same as the definition of $\bar{\lambda}_T$ for flexural buckling, except that now the elastic critical buckling force is that for torsional–flexural buckling (with the proviso that this is less than that for torsional buckling). Formulae for determining $N_{cr,T}$ and $N_{cr,TF}$ are not provided in EN 1993-1-1, but may be found in Part 1.3 of the code, and in Section 13.7 of this guide. Buckling curves for torsional and torsional–flexural buckling may be selected on the basis of Table 6.5 (*Table 6.2* of EN 1993-1-1), and by assuming buckling to be about the minor (z–z) axis.

Buckling lengths L_{cr}

Comprehensive guidance on buckling lengths for compression members with different end conditions is not provided in Eurocode 3, partly because no common consensus between the contributing countries could be reached. Some guidance on buckling lengths for compression members in triangulated and lattice structures is given in *Annex BB* of Eurocode 3. The provisions of *Annex BB* are discussed in Chapter 11 of this guide.

Typically, UK designers have been uncomfortable with the assumption of fully fixed end conditions, on the basis that there is inevitably a degree of flexibility in the connections. BS 5950: Part 1 therefore generally offers effective (or buckling) lengths that are less optimistic than the theoretical values. In the absence of Eurocode 3 guidance, it is therefore recommended that the BS 5950 buckling lengths be adopted. Table 6.6 contains the buckling lengths provided in clause 4.7.3 of BS 5950: Part 1; these buckling lengths are not to be applied to angles, channels or T-sections, for which reference should be made to clause 4.7.10 of BS 5950: Part 1. The boundary conditions and corresponding buckling lengths are illustrated in Figure 6.19, where L is equal to the system length. Further guidance is given in Brettle and Brown (2009).

Table 6.6. Nominal buckling lengths L_{cr} for compression members

End restraint (in the plane under consideration)		Buckling length, L_{cr}
Effectively held in position at both ends	Effectively restrained in direction at both ends	0.7L
	Partially restrained in direction at both ends	0.85L
	Restrained in direction at one end	0.85L
	Not restrained in direction at either end	1.0L

One end	Other end		Buckling length, L_{cr}
Effectively held in position and restrained in direction	Not held in position	Effectively restrained in direction	1.2L
		Partially restrained in direction	1.5L
		Not restrained in direction	2.0L

Figure 6.19. Nominal buckling lengths L_{cr} for compression members

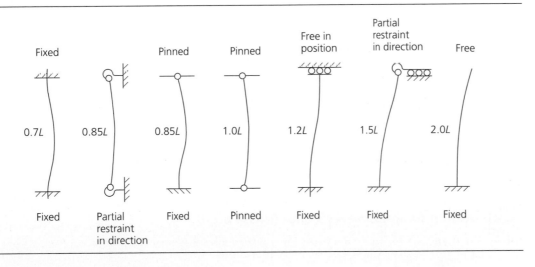

Example 6.7: buckling resistance of a compression member

A circular hollow section (CHS) member is to be used as an internal column in a multi-storey building. The column has pinned boundary conditions at each end, and the inter-storey height is 4 m, as shown in Figure 6.20. The critical combination of actions results in a design axial force of 2110 kN. Assess the suitability of a hot-rolled 244.5 × 10 CHS in grade S355 steel for this application.

Figure 6.20. General arrangement and loading

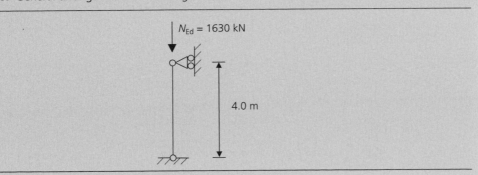

Section properties

The section properties are given in Figure 6.21.

Figure 6.21. Section properties for 244.5 × 10 CHS

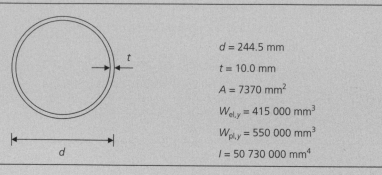

$d = 244.5$ mm

$t = 10.0$ mm

$A = 7370$ mm^2

$W_{el,y} = 415\,000$ mm^3

$W_{pl,y} = 550\,000$ mm^3

$I = 50\,730\,000$ mm^4

For a nominal material thickness ($t = 10.0$ mm) of less than or equal to 16 mm the nominal value of yield strength f_y for grade S355 steel is found from EN 10210-1 to be 355 N/mm^2.

Clause 3.2.6

From **clause 3.2.6**:

$E = 210\,000$ N/mm^2

Clause 5.5.2

Cross-section classification (clause 5.5.2)

$$\varepsilon = \sqrt{235/f_y} = \sqrt{235/355} = 0.81$$

Tubular sections (*Table 5.2*, sheet 3):

$$d/t = 244.5/10.0 = 24.5$$

Limit for Class 1 section $= 50\varepsilon^2 = 40.7$

$40.7 > 24.5$ ∴ section is Class 1

Clause 6.2.4

Cross-section compression resistance (clause 6.2.4)

$$N_{c,Rd} = \frac{Af_y}{\gamma_{M0}} \qquad \text{for Class 1, 2 or 3 cross-sections} \tag{6.10}$$

$$\therefore N_{c,Rd} = \frac{7370 \times 355}{1.00} = 2616 \times 10^3 \, \text{N} = 2616 \, \text{kN}$$

$2616 > 2110$ kN   ∴ cross-section resistance is acceptable

Clause 6.3.1

Member buckling resistance in compression (clause 6.3.1)

$$N_{b,Rd} = \frac{\chi Af_y}{\gamma_{M1}} \qquad \text{for Class 1, 2 and 3 cross-sections} \tag{6.47}$$

$$\chi = \frac{1}{\Phi + \sqrt{\Phi^2 - \bar{\lambda}^2}} \qquad \text{but } \chi \leq 1.0 \tag{6.49}$$

where

$$\Phi = 0.5[1 + \alpha(\bar{\lambda} - 0.2) + \bar{\lambda}^2]$$

and

$$\bar{\lambda} = \sqrt{\frac{Af_y}{N_{cr}}} \qquad \text{for Class 1, 2 and 3 cross-sections}$$

Elastic critical force and non-dimensional slenderness for flexural buckling

$$N_{cr} = \frac{\pi^2 EI}{l_u^2} = \frac{\pi^2 \times 210\,000 \times 50\,730\,000}{1000^2} = 6571\,kN$$

$$\therefore \bar{\lambda} = \sqrt{\frac{7370 \times 355}{6571 \times 10^3}} = 0.63$$

Selection of buckling curve and imperfection factor α

For a hot-rolled CHS, use buckling curve a (Table 6.5 (*Table 6.2* of EN 1993-1-1)).

For buckling curve a, $\alpha = 0.21$ (Table 6.4 (*Table 6.1* of EN 1993-1-1)).

Buckling curves

$$\Phi = 0.5[1 + 0.21 \times (0.63 - 0.2) + 0.63^2] = 0.74$$

$$\chi = \frac{1}{0.74 + \sqrt{0.74^2 - 0.63^2}} = 0.88$$

$$\therefore N_{b,Rd} = \frac{0.88 \times 7370 \times 355}{1.0} = 2297 \times 10^3\,N = 2297\,kN$$

$2297 > 2110\,kN \qquad \therefore$ buckling resistance is acceptable.

Conclusion

The chosen cross-section, 244.5×10 CHS, in grade S355 steel is acceptable.

6.3.2 Uniform members in bending

General

Laterally unrestrained beams subjected to bending about their major axis have to be checked for lateral torsional buckling (as well as for cross-sectional resistance), in accordance with *clause 6.3.2*. As described in Section 6.2.5 of this guide, there are a number of common situations where lateral torsional buckling need not be considered, and member strengths may be assessed on the basis of the in-plane cross-sectional strength.

Clause 6.3.2

EN 1993-1-1 contains three methods for checking the lateral torsional stability of a structural member:

- The primary method adopts the lateral torsional buckling curves given by *equations (6.56)* and *(6.57)*, and is set out in *clause 6.3.2.2* (general case) and *clause 6.3.2.3* (for rolled sections and equivalent welded sections). This method is discussed later in this section of the guide and illustrated in Example 6.8.
- The second is a simplified assessment method for beams with restraints in buildings, and is set out in *clause 6.3.2.4*. This method is discussed later in this section of the guide.
- The third is a general method for lateral and lateral torsional buckling of structural components, given in *clause 6.3.4* and discussed in the corresponding section of this guide.

Clause 6.3.2.2
Clause 6.3.2.3

Clause 6.3.2.4

Clause 6.3.4

A key aspect of designing laterally unrestrained beams is the determination of their non-dimensional lateral torsional buckling slenderness $\bar{\lambda}_{LT}$. As defined in EN 1993-1-1, this first requires calculation of the elastic buckling moment of the beam M_{cr}, though a more direct and simplified method for determining $\bar{\lambda}_{LT}$ is also available in NCCI SN002 (SCI, 2005a). Both methods are described later in this section of the guide.

Lateral restraint

Clause 6.3.2.1(2) deems that '*beams with sufficient lateral restraint to the compression flange are not susceptible to lateral torsional buckling*'. EN 1993-1-1 covers various cases of lateral restraint

Clause 6.3.2.1(2)

Clause 5.3.3(2) and imposes particular conditions to ensure their effectiveness. In general, a bracing system must be capable of resisting an equivalent stabilising force q_d (defined in *clause 5.3.3(2)*), the value of which depends on the flexibility of the bracing system. The design method is strictly iterative in nature, but a useful approach is to first assume the deflection of the bracing system, then to determine the resulting bracing forces, and finally to check that the assumed deflection is not exceeded. Assuming a deflection of L/2000, which will typically be conservative for bracing systems in buildings, results in bracing forces of 2% of the design force in the compression flange of the beam to be restrained.

Clause 6.3.5.2 Requirements for lateral restraints at plastic hinges are set out in *clause 6.3.5.2*, while lateral restraint from sheeting is covered in *Annex BB*.

Detailed guidance on all forms of lateral restraint is provided in Gardner (2011).

Lateral torsional buckling resistance

The design bending moment is denoted by M_{Ed} (bending moment design effect), and the lateral torsional buckling resistance by $M_{b,Rd}$ (design buckling resistance moment). Clearly, M_{Ed} must be shown to be less than $M_{b,Rd}$, and checks should be carried out on all unrestrained segments of beams (between the points where lateral restraint exists).

The design buckling resistance of a laterally unrestrained beam (or segment of beam) should be taken as

$$M_{b,Rd} = \chi_{LT} W_y \frac{f_y}{\gamma_{M1}} \tag{6.55}$$

where W_y is the section modulus appropriate for the classification of the cross-section, as given below. In determining W_y, no account need be taken for fastener holes at the beam ends.

$W_y = W_{pl,y}$ for Class 1 or 2 cross-sections

$W_y = W_{el,y}$ for Class 3 cross-sections

$W_y = W_{eff,y}$ for Class 4 cross-sections

χ_{LT} is the reduction factor for lateral torsional buckling.

From *equation (6.55)*, a clear analogy between the treatment of the buckling of bending members and the buckling of compression members can be seen. In both cases, the buckling resistance comprises a reduction factor (χ for compression; χ_{LT} for bending) multiplied by the cross-section strength ($A f_y / \gamma_{M1}$ for compression; $W_y f_y / \gamma_{M1}$ for bending).

Lateral torsional buckling curves

The lateral torsional buckling curves defined by EN 1993-1-1 are equivalent to (but not the same as) those set out in BS 5950: Part 1 in tabular form in Tables 16 and 17. Eurocode 3 provides four lateral torsional buckling curves (selected on the basis of the overall height-to-width ratio of the cross-section, the type of cross-section and whether the cross-section is rolled or welded), whereas BS 5950 offers only two curves (only making a distinction between rolled and welded sections).

Eurocode 3 defines lateral torsional buckling curves for two cases:

Clause 6.3.2.2
Clause 6.3.2.3
■ the general case (*clause 6.3.2.2*)
■ rolled sections or equivalent welded sections (*clause 6.3.2.3*).

Clause 6.3.2.2
Clause 6.3.2.3
Clause 6.3.2.2, the general case, may be applied to all common section types, including rolled sections, but, unlike *clause 6.3.2.3*, it may also be applied outside the standard range of rolled sections. For example, it may be applied to plate girders (of larger dimensions than standard rolled sections) and to castellated and cellular beams.

Table 6.7. Imperfection factors for lateral torsional buckling curves (*Table 6.3* of EN 1993-1-1)

Buckling curve	a	b	c	d
Imperfection factor α_{LT}	0.21	0.34	0.49	0.76

Lateral torsional buckling curves for the general case (*clause 6.3.2.2*) are described through *equation* (6.56):

$$\chi_{LT} = \frac{1}{\Phi_{LT} + \sqrt{\Phi_{LT}^2 - \bar\lambda_{LT}^2}} \quad \text{but } \chi_{LT} \le 1.0 \tag{6.56}$$

Clause 6.3.2.2

where

$$\Phi_{LT} = 0.5[1 + \alpha_{LT}(\bar\lambda_{LT} - 0.2) + \bar\lambda_{LT}^2]$$

$$\bar\lambda_{LT} = \sqrt{\frac{W_y f_y}{M_{cr}}}$$

α_{LT} is an imperfection factor from Table 6.7 (*Table 6.3* of EN 1993-1-1)
M_{cr} is the elastic critical moment for lateral torsional buckling (see the following subsection).

The imperfection factors α_{LT} for the four lateral torsional buckling curves are given by Table 6.7 (*Table 6.3* of EN 1993-1-1). Selection of the appropriate lateral torsional buckling curve for a given cross-section type and dimensions may be made with reference to Table 6.8 (*Table 6.4* of EN 1993-1-1).

Note that, although the 'default' plateau length (i.e. the non-dimensional slenderness below which the full in-plane bending moment resistance may be achieved) is set at 0.2 in *clause 6.3.2.2(1)*, a concession that allows lateral torsional buckling effects to be ignored up to a slenderness of $\bar\lambda_{LT,0}$ is made in *clause 6.3.2.2(4)*. $\bar\lambda_{LT,0}$ is defined in *clause 6.3.2.3*, and its value is given in *clause NA.2.17* of the UK National Annex as 0.4 for all rolled sections, including hollow sections, which generates a step in the buckling curves at this value. For welded sections, $\bar\lambda_{LT,0}$ is set at 0.2 in *clause NA.2.17*.

Clause 6.3.2.2(1)

Clause 6.3.2.2(4)
Clause 6.3.2.3
Clause NA.2.17
Clause NA.2.17

Lateral torsional buckling curves for the case of rolled sections or equivalent welded sections (*clause 6.3.2.3*) are described through *equation* (6.57) and also make use of the imperfection factors of Table 6.7 (*Table 6.3* of EN 1993-1-1). The definitions for $\bar\lambda_{LT}$, α_{LT} and M_{cr} are as for the general case, but the selection of lateral torsional buckling curve should be based on Table 6.9, which is the UK National Annex replacement for *Table 6.5* of EN 1993-1-1.

Clause 6.3.2.3

$$\chi_{LT} = \frac{1}{\Phi_{LT} + \sqrt{\Phi_{LT}^2 - \beta\bar\lambda_{LT}^2}} \quad \text{but } \chi_{LT} \le 1.0 \text{ and } \chi_{LT} \le \frac{1}{\bar\lambda_{LT}^2} \tag{6.57}$$

where

$$\Phi_{LT} = 0.5[1 + \alpha_{LT}(\bar\lambda_{LT} - \bar\lambda_{LT,0}) + \beta\bar\lambda_{LT}^2]$$

Table 6.8. Lateral torsional buckling curve selection table for the general case (*Table 6.4* of EN 1993-1-1)

Cross-section	Limits	Buckling curve
Rolled I sections	h/b ≤ 2 h/b > 2	a b
Welded I sections	h/b ≤ 2 h/b > 2	c d
Other cross-sections	–	d

Clause NA.2.17

Table 6.9. Lateral torsional buckling curve selection table for rolled or equivalent welded sections (replacement for *Table 6.5* of EN 1993-1-1 from *clause NA.2.17* of the UK National Annex)

Cross-section	Limits	Buckling curve
Rolled doubly symmetric I- and H-sections and hot-finished hollow sections	$h/b \leq 2$ $2.0 < h/b \leq 3.1$ $h/b > 3.1$	b c d
Angles (for moments in the major principal plane)		d
All other hot-rolled sections		d
Welded doubly symmetric sections and cold-formed hollow sections	$h/b \leq 2$ $2.0 \leq h/b < 3.1$	c d

Clause NA.2.17

The UK National Annex, through *clause NA.2.17*, defines $\bar{\lambda}_{LT,0} = 0.4$ and $\beta = 0.75$ for all rolled and hollow sections, and $\bar{\lambda}_{LT,0} = 0.2$ and $\beta = 1.0$ for all welded sections.

Clause 6.3.2.3

The method of *clause 6.3.2.3* also includes an additional factor f that is used to modify χ_{LT} (as shown by *equation (6.58)*)

$$\chi_{LT,mod} = \frac{\chi_{LT}}{f} \qquad \text{but } \chi_{LT,mod} \leq 1 \tag{6.58}$$

offering further enhancement in lateral torsional buckling resistance. Adopting $\chi_{LT,mod}$ is always beneficial, so could be safely ignored.

The factor f was derived on the basis of a numerical study, as

$$f = 1 - 0.5(1 - k_c)[1 - 2.0(\bar{\lambda}_{LT,0} - 0.8)^2] \tag{D6.8}$$

Clause NA.2.18

in which k_c is defined in *clause NA.2.18* of the UK National Annex as $k_c = 1/\sqrt{C_1}$, where C_1 is an equivalent uniform moment factor that depends on the shape of the bending moment diagram, and is discussed in the following section.

Clause 6.3.2.2
Clause 6.3.2.3

Figure 6.22 compares the lateral torsional buckling curves of the general case (*clause 6.3.2.2*) and the case for rolled sections or equivalent welded sections (*clause 6.3.2.3*). The imperfection factor α_{LT} for buckling curve b has been used for the comparison. Overall, it may be seen that the curve for the rolled or equivalent welded case is more favourable than that for the general case.

Elastic critical moment for lateral torsional buckling M_{cr}

As shown in the previous section, determination of the non-dimensional lateral torsional buckling slenderness $\bar{\lambda}_{LT}$ first requires calculation of the elastic critical moment for lateral torsional buckling M_{cr}. Eurocode 3 offers no formulations and gives no guidance on how M_{cr} should be calculated, except to say that M_{cr} should be based on gross cross-sectional properties and should take into account the loading conditions, the real moment distribution and the lateral

Clause 6.3.2.2(2)

restraints (*clause 6.3.2.2(2)*). Guidance is given, however, in NCCI SN002 (SCI, 2005a) and NCCI SN003 (SCI, 2005b), the key aspects of which are covered below. Note also, a simplified method for determining beam slenderness $\bar{\lambda}_{LT}$, which does not require calculation of M_{cr}, is provided in NCCI SN002 and discussed in the next section.

The elastic critical moment for lateral torsional buckling of a beam of uniform symmetrical cross-section with equal flanges, under standard conditions of restraint at each end, loaded through the shear centre and subject to uniform moment is given by equation (D6.9):

$$M_{cr,0} = \frac{\pi^2 E I_z}{L^2}\left(\frac{I_w}{I_z} + \frac{L^2 G I_T}{\pi^2 E I_z}\right)^{0.5} \tag{D6.9}$$

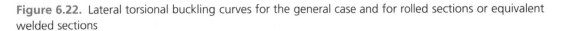

Figure 6.22. Lateral torsional buckling curves for the general case and for rolled sections or equivalent welded sections

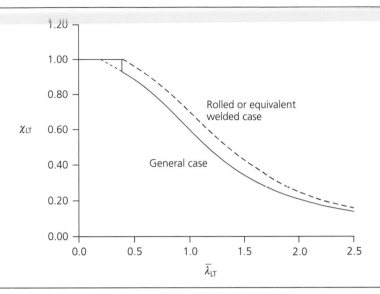

where

$$G = \frac{E}{2(1+v)}$$

I_T is the torsion constant
I_w is the warping constant
I_z is the second moment of area about the minor axis
L is the length of the beam between points of lateral restraint.

The standard conditions of restraint at each end of the beam are: restrained against lateral movement, restrained against rotation about the longitudinal axis and free to rotate on plan. Equation (D6.9) was provided in ENV 1993-1-1 (1992) in an informative Annex, and has been shown, for example by Timoshenko and Gere (1961), to represent the exact analytical solution to the governing differential equation.

Numerical solutions have also been calculated for a number of other loading conditions. For uniform doubly symmetric cross-sections, loaded through the shear centre at the level of the centroidal axis, and with the standard conditions of restraint described above, M_{cr} may be calculated through equation (D6.10):

$$M_{cr} = C_1 \frac{\pi^2 EI_z}{L^2} \left(\frac{I_w}{I_z} + \frac{L^2 GI_T}{\pi^2 EI_z} \right)^{0.5} \tag{D6.10}$$

where C_1 may be determined from Table 6.10 for end moment loading and from Table 6.11 for transverse loading. The C_1 factor is used to modify $M_{cr,0}$ (i.e. $M_{cr} = C_1 M_{cr,0}$) to take account of the shape of the bending moment diagram, and performs a similar function to the 'm' factor adopted in BS 5950. Note that values for the C_1 factor are derived numerically, and hence represent approximate solutions; for this reason, values from different sources (e.g. SCI, 2005a; Brettle and Brown, 2009) will vary slightly.

The values of C_1 given in Table 6.10 for end moment loading may be approximated by equation (D6.10), though other approximations also exist (Galambos, 1998):

$$C_1 = 1.88 - 1.40\psi + 0.52\psi^2 \qquad \text{but } C_1 \leq 2.70 \tag{D6.11}$$

where ψ is the ratio of the end moments (defined in Table 6.10).

Table 6.10. C_1 values for end moment loading

Loading and support conditions	Bending moment diagram		Value of C_1
		$\psi = +1$	1.000
		$\psi = +0.75$	1.141
		$\psi = +0.5$	1.323
		$\psi = +0.25$	1.563
		$\psi = 0$	1.879
		$\psi = -0.25$	2.281
		$\psi = -0.5$	2.704
		$\psi = -0.75$	2.927
		$\psi = -1$	2.752

Figure 6.23 compares values of C_1, which are obtained from Table 6.10 and from equation (D6.11). Figure 6.23 shows, as expected, that the most severe loading condition (that of uniform bending moment where $\psi = 1.0$) results in the lowest value for M_{cr}. As the ratio of the end moments ψ decreases, so the value of M_{cr} rises; these increases in M_{cr} are associated principally with changes that occur in the buckled deflected shape, which changes from a symmetric half sine wave for a uniform bending moment ($\psi = 1$) to an anti-symmetric double half wave for $\psi = -1$ (Trahair, 1993). At high values of C_1 there is some deviation between the approximate expression (equation (D6.11)) and the more accurate tabulated results of Table 6.10; thus, equation (D6.11) should not be applied when C_1 is greater than 2.70.

Table 6.11. C_1 values for transverse loading

Loading and support conditions	Bending moment diagram	Value of C_1
		1.132
		1.285
		1.365
		1.565
		1.046

Figure 6.23. Tabulated and approximate values of C_1 for varying ψ

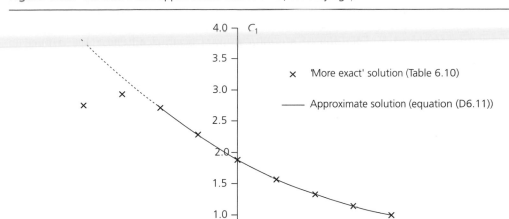

For differing degrees of end restraint against rotation on plan, equation (D6.10) is extended to

$$M_{cr} = C_1 \frac{\pi^2 EI_z}{(kL)^2} \left(\frac{I_w}{I_z} + \frac{(kL)^2 GI_T}{\pi^2 EI_z} \right) \tag{D6.12}$$

where k is an effective length parameter, values of which are given in Gardner (2011) and NCCI SN009 (SCI, 2005c).

A more general expression that allows for the shape of the bending moment diagram, different end restraint conditions, warping restraints, in-plane curvature prior to buckling, and the level at which the load is applied is given in NCCI SN002 (SCI, 2005a) (and without the allowance for in-plane curvature in NCCI SN003 (SCI, 2005b)) as

$$M_{cr} = C_1 \frac{\pi^2 EI_z}{(kL)^2 g} \left(\sqrt{ \left(\frac{k}{k_w} \right)^2 \frac{I_w}{I_z} + \frac{(kL)^2 GI_T}{\pi^2 EI_z} + (C_2 z_g)^2 } - C_2 z_g \right) \tag{D6.13}$$

where g allows for in-plane curvature of the beam prior to buckling, and is given by equation (D6.14), or may conservatively be taken as unity, k_w is a warping restraint parameter, z_g is the distance between the level of application of the loading and the shear centre (and is positive for destabilising loads applied above the shear centre) and C_2 is a parameter associated with the load level that is dependent on the shape of the bending moment diagram (SCI, 2005a; Gardner, 2011). Where no warping restraint is provided, and as a conservative assumption when the degree of warping restraint is uncertain, k_w should be taken equal to unity.

$$g = \sqrt{1 - \frac{I_z}{I_y}} \tag{D6.14}$$

Simplified determination of slenderness $\bar{\lambda}_{LT}$

The basic definition of non-dimensional beam slenderness $\bar{\lambda}_{LT}$ (defined in *clause 6.3.2.2*) requires the explicit calculation of M_{cr}, given, for the most general case by equation (D6.13) and, for more straightforward cases, by equations (D6.12) and (D6.10). Use of this approach will generally lead to the most accurate assessment of lateral torsional buckling resistance, and hence the most economic design. There are, however, a number of simplifications that can be made in the determination of $\bar{\lambda}_{LT}$ that will greatly expedite the calculation process, often with little loss of

Clause 6.3.2.2

economy. These simplifications are described in NCCI SN002 (SCI, 2005a), and are summarised below. A number of the simplifications relate specifically to doubly-symmetric I-sections.

Clause 6.3.2.2

As an alternative to the basic definition of non-dimensional beam slenderness $\bar{\lambda}_{\mathrm{LT}}$ defined in *clause 6.3.2.2*, $\bar{\lambda}_{\mathrm{LT}}$ may be determined from:

$$\bar{\lambda}_{\mathrm{LT}} = \frac{1}{\sqrt{C_1}} UVD\bar{\lambda}_z \sqrt{\beta_{\mathrm{w}}} \tag{D6.15}$$

in which C_1 is the equivalent uniform moment factor described previously, and U is a parameter that depends on section geometry, and is given in full below, where all symbols are as previously defined:

$$U = \sqrt{\frac{W_{\mathrm{pl},y}g}{A}} \sqrt{\frac{I_z}{I_{\mathrm{w}}}} \tag{D6.16}$$

For UB and UC sections, values of U range between about 0.84 and 0.90; $U = 0.9$ is therefore a suitable conservative upper bound for such sections. Tabulated values of U, calculated from equation (D6.16), are provided for standard hot-rolled steel sections in the SCI's 'Blue Book' (SCI/BCSA, 2009).

V is a parameter related to slenderness, and is given in full by

$$V = \frac{1}{\sqrt[4]{\left(\dfrac{k}{k_{\mathrm{w}}}\right)^2 + \dfrac{\lambda_z^2}{(\pi^2 E/G)(A/I_{\mathrm{T}})(I_{\mathrm{w}}/I_z)} + (C_2 z_{\mathrm{g}})^2 \dfrac{I_z}{I_{\mathrm{w}}}}} \tag{D6.17}$$

For doubly-symmetric hot-rolled UB and UC sections, and for cases where the loading is not destabilising, V may be conservatively simplified to:

$$V = \frac{1}{\sqrt[4]{1 + \dfrac{1}{20}\left(\dfrac{\lambda_z}{h/t_{\mathrm{f}}}\right)^2}} \tag{D6.18}$$

For all sections symmetric about the major axis and not subjected to destabilising loading, V may be conservatively taken as unity.

D is a destabilising parameter, given by

$$D = \frac{1}{\sqrt{1 - V^2 C_2 z_{\mathrm{g}} \sqrt{\dfrac{I_z}{I_{\mathrm{w}}}}}} \tag{D6.19}$$

to allow for destabilising loads (i.e. loads applied above the shear centre of the beam, where the load can move with the beam as it buckles). For loads applied at the level of the top flange, $D = 1.2$ is conservative. For non-destabilising loads, $D = 1.0$.

$\bar{\lambda}_z = \lambda_z/\lambda_1$ is the minor axis non-dimensional slenderness of the member, in which $\lambda_z = kL/i_z$, where k is the effective length parameter.

β_{w} is a parameter that allows for the classification of the cross-section: for Class 1 and 2 sections, $\beta_{\mathrm{w}} = 1$, while for Class 3 sections, $\beta_{\mathrm{w}} = W_{\mathrm{el},y}/W_{\mathrm{pl},y}$.

For a hot-rolled doubly-symmetric I- or H-section with lateral restraints to the compression flange at both ends of the segment under consideration and with no destabilising loads, non-dimensional beam slenderness $\bar{\lambda}_{\mathrm{LT}}$ may be conservatively obtained from Table 6.12, which has

Table 6.12. $\bar{\lambda}_{LT}$ for different steel grades (and yield strengths)

S235		S275		S355	
$f_y = 235$ N/mm²	$f_y = 225$ N/mm²	$f_y = 275$ N/mm²	$f_y = 265$ N/mm²	$f_y = 355$ N/mm²	$f_y = 345$ N/mm²
$\bar{\lambda}_{LT} = \dfrac{L/i_z}{104}$	$\bar{\lambda}_{LT} = \dfrac{L/i_z}{107}$	$\bar{\lambda}_{LT} = \dfrac{L/i_z}{96}$	$\bar{\lambda}_{LT} = \dfrac{L/i_z}{98}$	$\bar{\lambda}_{LT} = \dfrac{L/i_z}{85}$	$\bar{\lambda}_{LT} = \dfrac{L/i_z}{86}$

been derived on the basis of equation (D6.15) with the conservative assumptions of $C_1 = 1.0$, $U = 0.9$, $V = 1.0$, $D = 1.0$, $k = 1.0$ and $\sqrt{\beta_w} = 1$. Note that $\bar{\lambda}_{LT}$ is yield strength dependent, so for each steel grade, expressions for $\bar{\lambda}_{LT}$ have been provided for the two yield strengths associated with the two lowest thickness ranges: $t \leq 16$ mm and 16 mm $< t \leq 40$ mm.

The results obtained from Table 6.12 will be safe, but may become overly conservative when the shape of the bending moment diagram deviates significantly from uniform. In such circumstances, the influence of the shape of the bending moment diagram can be reintroduced simply by multiplying the expressions given in Table 6.12 by $1/\sqrt{C_1}$, in a similar fashion to equation (D6.15).

Example 6.8: lateral torsional buckling resistance

A simply supported primary beam is required to span 10.8 m and to support two secondary beams as shown in Figure 6.24. The secondary beams are connected through fin plates to the web of the primary beam, and full lateral restraint may be assumed at these points. Select a suitable member for the primary beam assuming grade S275 steel.

Figure 6.24. General arrangement

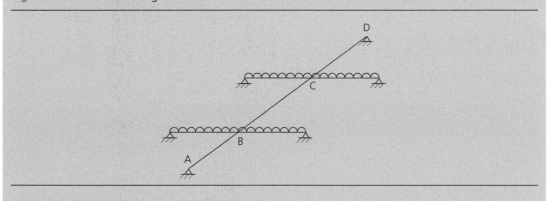

The loading, shear force and bending moment diagrams for the arrangement of Figure 6.24 are shown in Figure 6.25.

For the purposes of this worked example, lateral torsional buckling curves for the general case (*clause 6.3.2.2*) will be utilised.

Clause 6.3.2.2

Lateral torsional buckling checks will be carried out on segments BC and CD. By inspection, segment AB is not critical.

Consider a 762 × 267 × 173 UKB in grade S275 steel.

Section properties
The section properties are shown in Figure 6.26.

For a nominal material thickness ($t_f = 21.6$ mm and $t_w = 14.3$ mm) of greater than 16 mm but less than or equal to 40 mm, the nominal value of yield strength f_y for grade S275 steel is found from EN 10025-2 to be 265 N/mm².

Figure 6.25. (a) Loading, (b) shear forces and (c) bending moments

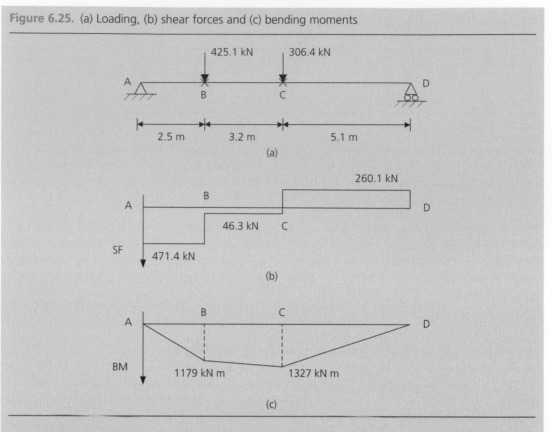

From *clause 3.2.6*:

$$E = 210\,000 \text{ N/mm}^2$$

$$G \approx 81\,000 \text{ N/mm}^2$$

Cross-section classification (*clause 5.5.2*)

$$\varepsilon = \sqrt{235/f_y} = \sqrt{235/265} = 0.94$$

Outstand flanges (*Table 5.2*, sheet 2):

$$c_f = (b - t_w - 2r)/2 = 109.7 \text{ mm}$$

$$c_f/t_f = 109.7/21.6 = 5.08$$

Figure 6.26. Section properties for a $762 \times 267 \times 173$ UKB

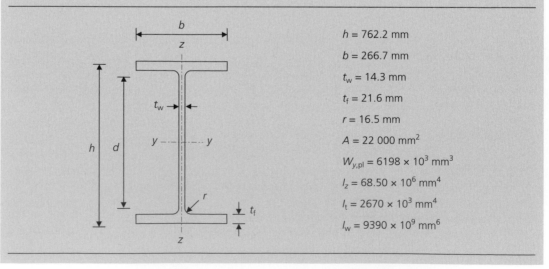

$h = 762.2$ mm

$b = 266.7$ mm

$t_w = 14.3$ mm

$t_f = 21.6$ mm

$r = 16.5$ mm

$A = 22\,000$ mm^2

$W_{y,pl} = 6198 \times 10^3$ mm^3

$I_z = 68.50 \times 10^6$ mm^4

$I_t = 2670 \times 10^3$ mm^4

$I_w = 9390 \times 10^9$ mm^6

Limit for Class 1 flange $= 9\varepsilon = 8.48$

$8.48 > 5.08 \qquad \therefore$ flange is Class 1

Web – internal part in bending (*Table 5.2*, sheet 1):

$c_w = h - 2t_f - 2r = 686.0$ mm

$c_w/t_w = 686.0/14.3 = 48.0$

Limit for Class 1 web $= 72\varepsilon = 67.8$

$67.8 > 48.0 \qquad \therefore$ web is Class 1

The overall cross-section classification is therefore Class 1.

Bending resistance of cross-section (clause 6.2.5)

Clause 6.2.5

$$M_{c,y,Rd} = \frac{W_{pl,y}f_y}{\gamma_{M0}} \qquad \text{for Class 1 or 2 cross-sections} \qquad (6.13)$$

From *clause NA.2.15* of the UK National Annex, $\gamma_{M0} = 1.00$.

Clause NA.2.15

The design bending resistance of the cross-section

$$M_{c,y,Rd} = \frac{6198 \times 10^3 \times 265}{1.00} = 1642 \times 10^6 \, \text{N mm} = 1642 \, \text{kN m}$$

$1642 \, \text{kN m} > 1327 \, \text{kN m} \qquad \therefore$ cross-section resistance in bending is acceptable

Shear resistance of cross-section (clause 6.2.6)

Clause 6.2.6

$$V_{pl,Rd} = \frac{A_v(f_y/\sqrt{3})}{\gamma_{M0}} \qquad (6.18)$$

For a rolled I-section, loaded parallel to the web, the shear area A_v is given by

$A_v = A - 2bt_f + (t_w + 2r)t_f \qquad$ (but not less than $\eta h_w t_w$)

From *clause NA.2.4* of the UK National Annex to EN 1993-1-5, $\eta = 1.0$.

Clause NA.2.4

$h_w = h - 2t_f = 762.2 - (2 \times 21.6) = 719.0$ mm

$\therefore A_v = 22\,000 - (2 \times 266.7 \times 21.6) + (14.3 + [2 \times 16.5]) \times 21.6$

$\qquad = 11\,500 \, \text{mm}^2 \qquad$ (but not less than $1.0 \times 719.0 \times 14.3 = 10\,282 \, \text{mm}^2$)

$\therefore V_{pl,Rd} = \dfrac{11\,500 \times (265/\sqrt{3})}{1.00} = 1\,759\,000 \, \text{N} = 1759 \, \text{kN}$

Shear buckling need not be considered provided

$$\frac{h_w}{t_w} \le 72\frac{\varepsilon}{\eta} \qquad \text{for unstiffened webs}$$

$72\dfrac{\varepsilon}{\eta} = 72 \times \dfrac{0.94}{1.0} = 67.8$

Actual $h_w/t_w = 719.0/14.3 = 50.3$

$50.3 \le 67.8 \qquad \therefore$ no shear buckling check required

$1759 > 471.4 \, \text{kN} \qquad \therefore$ shear resistance is acceptable

Resistance of cross-section under combined bending and shear (clause 6.2.8)

Clause 6.2.8
Clause 6.2.8

Clause 6.2.8 states that provided the shear force V_{Ed} is less than half the plastic shear resistance $V_{pl,Rd}$ its effect on the moment resistance may be neglected except where shear buckling reduces the section resistance. In this case, there is no reduction for shear buckling (see above), and the maximum shear force ($V_{Ed} = 471.4\,\text{kN}$) is less than half the plastic shear resistance ($V_{pl,Rd} = 1759\,\text{kN}$). Therefore, resistance under combined bending and shear is acceptable.

Clause 6.3.2.2

Lateral torsional buckling check (clause 6.3.2.2): segment BC

$$M_{Ed} = 1327\,\text{kN m}$$

$$M_{b,Rd} = \chi_{LT} W_y \frac{f_y}{\gamma_{M1}} \tag{6.55}$$

where

$$W_y = W_{pl,y} \text{ for Class 1 and 2 cross-sections}$$

Determine M_{cr}: segment BC ($L = 3200$ mm)
The elastic buckling moment M_{cr} is determined from

$$M_{cr} = C_1 \frac{\pi^2 EI_z}{L^2} \left(\frac{I_w}{I_z} + \frac{L^2 GI_T}{\pi^2 EI_z}\right)^{0.5} \tag{D6.10}$$

since the load is not destabilising and the effective length parameter is assumed to be unity.

Approximate C_1 from equation (D6.11):

$$C_1 = 1.88 - 1.40\psi + 0.52\psi^2 \qquad (\text{but } C_1 \leq 2.7)$$

ψ is the ratio of end moments $= 1179/1327 = 0.89$

$$\Rightarrow C_1 = 1.05$$

$$\therefore M_{cr} = 1.05 \times \frac{\pi^2 \times 210\,000 \times 68.5 \times 10^6}{3200^2} \times \left(\frac{9390 \times 10^9}{68.5 \times 10^6} + \frac{3200^2 \times 81\,000 \times 2670 \times 10^3}{\pi^2 \times 210\,000 \times 68.5 \times 10^6}\right)^{0.5}$$

$$= 5670 \times 10^6\,\text{N mm} = 5670\,\text{kNm}$$

Non-dimensional lateral torsional slenderness $\bar{\lambda}_{LT}$: segment BC

$$\bar{\lambda}_{LT} = \sqrt{\frac{W_y f_y}{M_{cr}}} = \sqrt{\frac{6198 \times 10^3 \times 265}{5670 \times 10^6}} = 0.54$$

Select buckling curve and imperfection factor α_{LT}
Using Table 6.8 (*Table 6.4 of EN 1993-1-1*),

$$h/b = 762.2/266.7 = 2.85$$

Therefore, for a rolled I section with $h/b > 2$, use buckling curve b.

For buckling curve b, $\alpha_{LT} = 0.34$ from Table 6.7 (*Table 6.3 of EN 1993-1-1*).

Calculate reduction factor for lateral torsional buckling, χ_{LT}: segment BC

$$\chi_{LT} = \frac{1}{\Phi_{LT} + \sqrt{\Phi_{LT}^2 - \bar{\lambda}_{LT}^2}} \qquad \text{but } \chi_{LT} \leq 1.0 \tag{6.56}$$

where

$$\Phi_{LT} = [1 + \alpha_{LT}(\bar{\lambda}_{LT} - 0.2) + \bar{\lambda}_{LT}^2]$$

$$= 0.5 \times [1 + 0.34 \times (0.54 - 0.2) + 0.54^2] = 0.70$$

$$\therefore \chi_{LT} = \frac{1}{0.70 + \sqrt{0.70^2 - 0.54^2}} = 0.87$$

Lateral torsional buckling resistance: segment BC

$$M_{b,Rd} = \chi_{LT} W_y \frac{f_y}{\gamma_{M1}}$$ (6.55)

$$= 0.87 \times 6198 \times 10^3 \times (265/1.0)$$

$$= 1424 \times 10^6 \text{ N mm} = 1424 \text{ kN m}$$

$$\frac{M_{Ed}}{M_{b,Rd}} = \frac{1327}{1424} = 0.93$$

$0.93 \leq 1.0$ $\therefore$ segment BC is acceptable

Lateral torsional buckling check (clause 6.3.2.2): segment CD

Clause 6.3.2.2

$$M_{Ed} = 1327 \text{ kN m}$$

Determine M_{cr}: segment CD ($L = 5100$ mm)

$$M_{cr} = C_1 \frac{\pi^2 EI_z}{L^2} \left(\frac{I_w}{I_z} + \frac{L^2 GI_T}{\pi^2 EI_z} \right)^{0.5}$$ (D6.10)

Determine C_1 from Table 6.11 (or approximate from equation (D6.11)):

ψ is the ratio of end moments $= 0/1362 = 0$

$\Rightarrow C_1 = 1.879$ from Table 6.11

$$\therefore M_{cr} = 1.879 \times \frac{\pi^2 \times 210\,000 \times 68.5 \times 10^6}{5100^2}$$

$$\times \left(\frac{9390 \times 10^9}{68.5 \times 10^6} + \frac{5100^2 \times 81\,000 \times 2670 \times 10^3}{\pi^2 \times 210\,000 \times 68.5 \times 10^6} \right)^{0.5}$$

$$= 4311 \times 10^6 \text{ N mm} = 4311 \text{ kN m}$$

Non-dimensional lateral torsional slenderness $\bar{\lambda}_{LT}$: segment CD

$$\bar{\lambda}_{LT} = \sqrt{\frac{W_y f_y}{M_{cr}}} = \sqrt{\frac{6198 \times 10^3 \times 265}{4311 \times 10^6}} = 0.62$$

The buckling curve and imperfection factor α_{LT} are as for segment BC.

Calculate reduction factor for lateral torsional buckling, χ_{LT}: segment CD

$$\chi_{LT} = \frac{1}{\Phi_{LT} + \sqrt{\Phi_{LT}^2 - \bar{\lambda}_{LT}^2}} \qquad \text{but } \chi_{LT} \leq 1.0$$ (6.56)

where

$$\Phi_{LT} = 0.5[1 + \alpha_{LT}(\bar{\lambda}_{LT} - 0.2) + \bar{\lambda}_{LT}^2]$$

$$= 0.5 \times [1 + 0.34 \times (0.62 - 0.2) + 0.62^2] = 0.76$$

$$\therefore \chi_{LT} = \frac{1}{0.76 + \sqrt{0.76^2 - 0.62^2}} = 0.83$$

Lateral torsional buckling resistance: segment CD

From *clause NA.2.15* of the UK National Annex, $\gamma_{M1} = 1.00$.

$$M_{b,Rd} = \chi_{LT} W_y \frac{f_y}{\gamma_{M1}} \tag{6.55}$$

$$= 0.83 \times 6198 \times 10^3 \times (265/1.0)$$

$$= 1360 \times 10^6 \text{ N mm} = 1360 \text{ kN m}$$

$$\frac{M_{Ed}}{M_{b,Rd}} = \frac{1327}{1360} = 0.98$$

$$0.98 \leq 1.0 \qquad \therefore \text{ segment CD is acceptable}$$

Conclusion

The design is controlled by the lateral stability of segment CD. The chosen cross-section, $762 \times 267 \times 173$ UKB, in grade S275 steel is acceptable.

Simplified assessment methods for beams with restraints in buildings

Clause 6.3.2.4 provides a quick, approximate and conservative way of determining whether the lengths of a beam between points of effective lateral restraints L_c will be satisfactory under its maximum design moment $M_{y,Ed}$, expressed as a fraction of the resistance moment of the cross-section $M_{c,Rd}$. In determining $M_{c,Rd}$, the section modulus W_y must relate to the compression flange.

For the simplest case when the steel strength $f_y = 235 \text{ N/mm}^2$ (and thus $\varepsilon = 1.0$), $M_{y,Ed}$ is equal to $M_{c,Rd}$, and uniform moment loading is assumed, the condition reduces to

$$L_c \leq 37.6 i_{f,z} \tag{D6.20}$$

in which $i_{f,z}$ is the radius of gyration of the compression flange plus one-third of the compressed portion of the web, about the minor axis, and $\bar{\lambda}_{c0}$ has been taken as 0.4 from *clause NA.2.19* of the UK National Annex.

More generally, the limit may be expressed in the form of *equation (6.59)*:

$$\bar{\lambda}_f = \frac{k_c L_c}{i_{f,z} \lambda_1} \leq \bar{\lambda}_{c0} \frac{M_{c,Rd}}{M_{y,Rd}} \tag{6.59}$$

where $k_c = 1/\sqrt{C_1}$, from *clause NA.2.18* of the UK National Annex and allows for different patterns of moments between restraint points with C_1 obtained from Table 6.10 or 6.11, and $\lambda_1 = 93.9\varepsilon$.

Clearly, if the required level of moment $M_{y,Ed}$ is less than $M_{c,Rd}$, then the value of $\bar{\lambda}_f$, and hence L_c, will increase pro rata.

6.3.3 Uniform members in bending and axial compression

Members subjected to bi-axial bending and axial compression (beam–columns) exhibit complex structural behaviour. First-order bending moments about the major and minor axes ($M_{y,Ed}$ and $M_{z,Ed}$, respectively) are induced by lateral loading and/or end moments. The addition of axial loading N_{Ed} clearly results in axial force in the member, but also amplifies the bending moments about both principal axes (second-order bending moments). Since, in general, the bending moment distributions about both principal axes will be non-uniform (and hence the most heavily loaded cross-section can occur at any point along the length of the member), plus there is a coupling between the response in the two principal planes, design treatment is necessarily complex. The behaviour and design of beam–columns is covered thoroughly by Chen and Atsuta (1977).

Although there is a coupling between the member response in the two principal planes, this is generally safely disregarded in design. Instead, a pair of interaction equations, which essentially check member resistance about each of the principal axes (y–y and z–z) is employed. In *clause 6.3.3* such a pair of interaction equations is provided (see *equations (6.61)* and *(6.62)*) to check the resistance of individual lengths of members between restraints, subjected to known bending moments and axial forces. Both interaction equations must be satisfied. Second-order sway effects (P–Δ effects) should be allowed for, either by using suitably enhanced end moments or by using appropriate buckling lengths. It is also specifically noted that the cross-section resistance at each end of the member should be checked against the requirements of *clause 6.2*.

Clause 6.3.3

Clause 6.2

Two classes of problem are recognised:

- members not susceptible to torsional deformation
- members susceptible to torsional deformation.

The former is for cases where no lateral torsional buckling is possible, for example where square or circular hollow sections are employed, as well as arrangements where torsional deformation is prevented, such as open sections restrained against twisting. Most I- and H-section columns in building frames are likely to fall within the second category.

At first sight, *equations (6.61)* and *(6.62)* appear similar to the equations given in clause 4.8.3.3 of BS 5950: Part 1. However, determination of the interaction or k factors is significantly more complex. Omitting the terms required only to account for the shift in neutral axis (from the gross to the effective section) for Class 4 cross-sections, the formulae are

$$\frac{N_{Ed}}{\chi_y N_{Rk}/\gamma_{M1}} + k_{yy}\frac{M_{y,Ed}}{\chi_{LT}M_{y,Rk}/\gamma_{M1}} + k_{yz}\frac{M_{z,Ed}}{M_{z,Rk}/\gamma_{M1}} \leq 1 \qquad (6.61)$$

$$\frac{N_{Ed}}{\chi_z N_{Rk}/\gamma_{M1}} + k_{zy}\frac{M_{y,Ed}}{\chi_{LT}M_{y,Rk}/\gamma_{M1}} + k_{zz}\frac{M_{z,Ed}}{M_{z,Rk}/\gamma_{M1}} \leq 1 \qquad (6.62)$$

in which

N_{Ed}, $M_{y,Ed}$, $M_{z,Ed}$ are the design values of the compression force and the maximum moments about the y–y and z–z axes along the member, respectively

N_{Rk}, $M_{y,Rk}$, $M_{z,Rk}$ are the characteristic values of the compression resistance of the cross-section and the bending moment resistances of the cross-section about the y–y and z–z axes, respectively

χ_y, χ_z are the reduction factors due to flexural buckling from *clause 6.3.1*

χ_{LT} is the reduction factor due to lateral torsional buckling from *clause 6.3.2*, taken as unity for members that are not susceptible to torsional deformation

k_{yy}, k_{yz}, k_{zy}, k_{zz} are the interaction factors k_{ij}.

Clause 6.3.1
Clause 6.3.2

The characteristic values of the cross-sectional resistances N_{Rk}, $M_{y,Rk}$ and $M_{z,Rk}$ may be calculated as for the design resistances, but without dividing by the partial γ_M factor. The relationship between characteristic and design resistance is given by *equation (2.1)*.

Values for the interaction factors k_{ij} are to be obtained from one of two methods given in *Annex A* (alternative method 1) or *Annex B* (alternative method 2). These originate from two different approaches to the beam–column interaction problem – enhancing the elastic resistance, taking account of buckling effects to include partial plastification of the cross-section, or reducing the plastic cross-sectional resistance to allow for instability effects. Both approaches distinguish between cross-sections susceptible or not susceptible to torsion, as well as between elastic (for Class 3 and 4 cross-sections) and plastic (for Class 1 and 2 cross-sections) properties. The methods are discussed in more detail in Chapters 8 and 9 of this guide. The UK National Annex limits the scope of application of *Annex A* to bi-symmetrical sections, while the simpler

Table 6.13. Safe (maximum) values for interaction factors

Interaction factor	Class 1 and 2	Class 3
k_{yy}	$1.8C_{my}$	$1.6C_{my}$
k_{yz}	$0.6k_{zz}$	k_{zz}
k_{zy}	1.0	1.0
k_{zz}	$2.4C_{mz}$	$1.6C_{mz}$

Annex B may be applied in all cases, although with some restrictions for sections other than I-, H- or hollow sections.

Determination of the interaction factors can be a rather lengthy process, and, with many of the intermediate parameters lacking a clear physical meaning, can be prone to errors. Some form of automation, such as through the use of spreadsheets or programing, is recommended to aid the process. To facilitate the design process, Brettle and Brown (2009) provide (based on *Annex B*) both a graphical means for the accurate determination of interaction factors and a set of safe (maximum) values that may be used directly in *equations (6.61)* and *(6.62)*. These safe values, given in Table 6.13, offer speed and convenience for hand calculations, but will often be overly conservative.

Aside from determination of the interaction factors, all other calculations relate to the individual member checks under either compression or bending, described in the two previous sections of this guide.

Example 6.9 considers member resistance under combined major axis bending and axial load, and uses alternative method 1 (*Annex A*) to determine the necessary interaction factors k_{ij}.

Example 6.9: member resistance under combined major axis bending and axial compression

A rectangular hollow section (RHS) member is to be used as a primary floor beam of 7.2 m span in a multi-storey building. Two design point loads of 58 kN are applied to the primary beam (at locations B and C) from secondary beams, as shown in Figure 6.27. The secondary beams are connected through fin plates to the webs of the primary beam, and full lateral and torsional restraint may be assumed at these points. The primary beam is also subjected to a design axial force of 90 kN.

Assess the suitability of a hot-rolled $200 \times 100 \times 16$ RHS in grade S355 steel for this application.

In this example the interaction factors k_{ij} (for member checks under combined bending and axial compression) will be determined using alternative method 1 (*Annex A*), which is discussed in Chapter 8 of this guide.

Figure 6.27. General arrangement and loading

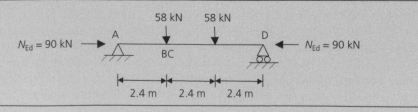

Section properties
The section properties are given in Figure 6.28.

Figure 6.28. Section properties for $200 \times 100 \times 16$ RHS

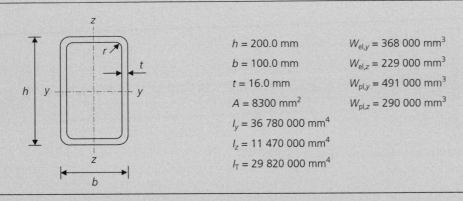

$h = 200.0$ mm $W_{el,y} = 368\,000$ mm^3

$b = 100.0$ mm $W_{el,z} = 229\,000$ mm^3

$t = 16.0$ mm $W_{pl,y} = 491\,000$ mm^3

$A = 8300$ mm^2 $W_{pl,z} = 290\,000$ mm^3

$I_y = 36\,780\,000$ mm^4

$I_z = 11\,470\,000$ mm^4

$I_T = 29\,820\,000$ mm^4

For a nominal material thickness ($t = 16.0$ mm) of less than or equal to 16 mm the nominal value of yield strength f_y for grade S355 steel is found from EN 10210-1 to be 355 N/mm^2.

From *clause 3.2.6*:

Clause 3.2.6

$$E = 210\,000 \text{ N/mm}^2$$

$$G \approx 81\,000 \text{ N/mm}^2$$

Cross-section classification (clause 5.5.2)

Clause 5.5.2

$$\varepsilon = \sqrt{235/f_y} = \sqrt{235/355} = 0.81$$

For a RHS the compression width c may be taken as h (or b) – $3t$.

Flange – internal part in compression (*Table 5.2*, sheet 1):

$$c_f = b - 3t = 100.0 - (3 \times 16.0) = 52.0 \text{ mm}$$

$$c_f/t = 52.0/16.0 = 3.25$$

Limit for Class 1 flange $= 33\varepsilon = 26.85$

$26.85 > 3.25$ $\therefore$ flange is Class 1

Web – internal part in compression (*Table 5.2*, sheet 1):

$$c_w = h - 3t = 200.0 - (3 \times 16.0) = 152.0 \text{ mm}$$

$$c_w/t = 152.0/16.0 = 9.50$$

Limit for Class 1 web $= 33\varepsilon = 26.85$

$26.85 > 9.50$ $\therefore$ web is Class 1

The overall cross-section classification is therefore Class 1 (under pure compression).

Compression resistance of cross-section (clause 6.2.4)
The design compression resistance of the cross-section $N_{c,Rd}$

Clause 6.2.4

$$N_{c,Rd} = \frac{Af_y}{\gamma_{M0}} \quad \text{for Class 1, 2 or 3 cross-sections} \tag{6.10}$$

$$= \frac{8300 \times 355}{1.00} = 2\,946\,500 \text{ N} = 2946.5 \text{ kN}$$

$2946.5 \text{ kN} > 90 \text{ kN}$ $\therefore$ acceptable

Clause 6.2.5

Bending resistance of cross-section (clause 6.2.5)
Maximum bending moment

$$M_{y,Ed} = 2.4 \times 58 = 139.2 \text{ kN m}$$

The design major axis bending resistance of the cross-section

$$M_{c,y,Rd} = \frac{W_{pl,y}f_y}{\gamma_{M0}} \qquad \text{for Class 1 or 2 cross-sections} \tag{6.13}$$

$$= \frac{491\,000 \times 355}{1.00} = 174.3 \times 10^6 \text{ N mm} = 174.3 \text{ kN m}$$

$$174.3 \text{ kN m} > 139.2 \text{ kN m} \qquad \therefore \text{ acceptable}$$

Clause 6.2.6

Shear resistance of cross-section (clause 6.2.6)
Maximum shear force

$$V_{Ed} = 58.0 \text{ kN}$$

The design plastic shear resistance of the cross-section

$$V_{pl,Rd} = \frac{A_v(f_y/\sqrt{3})}{\gamma_{M0}} \tag{6.18}$$

or a rolled RHS of uniform thickness, loaded parallel to the depth, the shear area A_v is given by

$$A_v = Ah/(b+h) = 8300 \times 200/(100+200) = 5533 \text{ mm}^2$$

$$\therefore V_{pl,Rd} = \frac{5533 \times (355/\sqrt{3})}{1.00} = 1134 \times 10^3 \text{ N} = 1134 \text{ kN}$$

Shear buckling need not be considered, provided

$$\frac{h_w}{t_w} \leq 72\frac{\varepsilon}{\eta} \qquad \text{for unstiffened webs}$$

Clause NA.2.4

$\eta = 1.0$ from *clause NA.2.4* of the UK National Annex to EN 1993-1-5.

$$h_w = (h - 2t) = 200 - (2 \times 16.0) = 168 \text{ mm}$$

$$72\frac{\varepsilon}{\eta} = 72 \times \frac{0.81}{1.0} = 58.6$$

Actual $h_w/t_w = 168/16.0 = 10.5$

$$10.5 \leq 58.6 \qquad \therefore \text{ no shear buckling check required}$$

$$1134 > 58.0 \text{ kN} \qquad \therefore \text{ shear resistance is acceptable}$$

Clause 6.2.10

Cross-section resistance under bending, shear and axial force (clause 6.2.10)
Provided the shear force V_{Ed} is less than 50% of the design plastic shear resistance $V_{pl,Rd}$,

Clause 6.2.9

and provided shear buckling is not a concern, then the cross-section need only satisfy the requirements for bending and axial force (*clause 6.2.9*).

In this case $V_{Ed} < 0.5V_{pl,Rd}$, and shear buckling is not a concern (see above). Therefore, cross-section only need be checked for bending and axial force.

No reduction to the major axis plastic resistance moment due to the effect of axial force is required when both of the following criteria are satisfied:

$$N_{Ed} \leq 0.25N_{pl,Rd} \tag{6.33}$$

$$N_{Ed} \leq \frac{0.5h_w t_w f_y}{\gamma_{M0}} \tag{6.34}$$

$0.25 N_{pl,Rd} = 0.25 \times 2946.5 = 736.6\,\text{kN}$

$736.6\,\text{kN} > 90\,\text{kN} \qquad \therefore \text{ equation (6.33) is satisfied}$

$\dfrac{0.5 h_w t_w f_y}{\gamma_{M0}} = \dfrac{0.5 \times 168.0 \times (2 \times 16.0) \times 355}{1.0} = 954.2\,\text{kN}$

$954.2\,\text{kN} > 90\,\text{kN} \qquad \therefore \text{ equation (6.34) is satisfied}$

Therefore, no allowance for the effect of axial force on the major axis plastic moment resistance of the cross-section need be made.

Member buckling resistance in compression (clause 6.3.1)

Clause 6.3.1

$$N_{b,Rd} = \frac{\chi A f_y}{\gamma_{M1}} \qquad \text{for Class 1, 2 and 3 cross-sections} \tag{6.47}$$

$$\chi = \frac{1}{\Phi + \sqrt{\Phi^2 - \bar{\lambda}^2}} \qquad \text{but } \chi \leq 1.0 \tag{6.49}$$

where

$$\Phi = 0.5[1 + \alpha(\bar{\lambda} - 0.2) + \bar{\lambda}^2]$$

$$\bar{\lambda} = \sqrt{\frac{A f_y}{N_{cr}}} \qquad \text{for Class 1, 2 and 3 cross-sections}$$

Elastic critical force and non-dimensional slenderness for flexural buckling

For buckling about the major (y–y) axis, L_{cr} should be taken as the full length of the beam (AD), which is 7.2 m. For buckling about the minor (z–z) axis, L_{cr} should be taken as the maximum length between points of lateral restraint, which is 2.4 m. Thus,

$$N_{cr,y} = \frac{\pi^2 E I_y}{L_{cr}^2} = \frac{\pi^2 \times 210\,000 \times 36\,780\,000}{7200^2} = 1470 \times 10^3\,\text{N} = 1470\,\text{kN}$$

$$\therefore \bar{\lambda}_y = \sqrt{\frac{8300 \times 355}{1470 \times 10^3}} = 1.42$$

$$N_{cr,z} = \frac{\pi^2 E I_z}{L_{cr}^2} = \frac{\pi^2 \times 210\,000 \times 11\,470\,000}{2400^2} = 4127 \times 10^3\,\text{N} = 4127\,\text{kN}$$

$$\therefore \bar{\lambda}_z = \sqrt{\frac{8300 \times 355}{4127 \times 10^3}} = 0.84$$

Selection of buckling curve and imperfection factor α

For a hot-rolled RHS, use buckling curve a (Table 6.5 (*Table 6.2* of EN 1993-1-1)).

For buckling curve *a*, $\alpha = 0.21$ (Table 6.4 (*Table 6.1* of EN 1993-1-1)).

Buckling curves: major (y–y) axis

$$\Phi_y = 0.5 \times [1 + 0.21 \times (1.42 - 0.2) + 1.42^2] = 1.63$$

$$\chi_y = \frac{1}{1.63 + \sqrt{1.63^2 - 1.42^2}} = 0.41$$

$$\therefore N_{b,y,Rd} = \frac{0.41 \times 8300 \times 355}{1.0} = 1209 \times 10^3\,\text{N} = 1209\,\text{kN}$$

$1209\,\text{kN} > 90\,\text{kN} \qquad \therefore \text{ major axis flexural buckling resistance is acceptable}$

Buckling curves: minor (z–z) axis

$$\Phi_z = 0.5 \times [1 + 0.21 \times (0.84 - 0.2) + 0.84^2] = 0.92$$

$$\chi_z = \frac{1}{0.92 + \sqrt{0.92^2 - 0.84^2}} = 0.77$$

$$\therefore N_{b,z,Rd} = \frac{0.77 \times 8300 \times 355}{1.0} = 2266 \times 10^3 \, \text{N} = 2266 \, \text{kN}$$

$2266 \, \text{kN} > 90 \, \text{kN}$ $\therefore$ minor axis flexural buckling resistance is acceptable

Clause 6.3.2

Member buckling resistance in bending (clause 6.3.2)

By inspection, the central segment (BC) of the beam is critical (since it is subjected to uniform bending and of equal length to the two outer segments). Therefore, only segment BC need be checked.

$$M_{Ed} = 139.2 \, \text{kN m}$$

$$M_{b,Rd} = \chi_{LT} W_y \frac{f_y}{\gamma_{M1}} \tag{6.55}$$

where

$W_y = W_{pl,y}$ for Class 1 and 2 cross-sections

Determine M_{cr} ($kL_{cr} = 2400$ mm)

$$M_{cr} = C_1 \frac{\pi^2 E I_z}{kL^2} \left(\frac{I_w}{I_z} + \frac{kL^2 G I_T}{\pi^2 E I_z} \right)^{0.5} \tag{D6.12}$$

For a uniform bending moment, $C_1 = 1.0$ (from Table 6.12).

Since the cross-section is closed, the warping contribution will be ignored.

$$\therefore M_{cr} = 1.0 \times \frac{\pi^2 \times 210\,000 \times 11.47 \times 10^6}{2400^2} \left(\frac{2400^2 \times 81\,000 \times 29.82 \times 10^6}{\pi^2 \times 210\,000 \times 11.47 \times 10^6} \right)^{0.5}$$

$$= 3157 \times 10^6 \, \text{N mm} = 3157 \, \text{kN m}$$

Non-dimensional lateral torsional slenderness $\bar{\lambda}_{LT}$: segment BC

$$\bar{\lambda}_{LT} = \sqrt{\frac{W_y f_y}{M_{cr}}} = \sqrt{\frac{491 \times 10^3 \times 355}{3157 \times 10^6}} = 0.23$$

$$\bar{\lambda} \leq \bar{\lambda}_{LT,0} = 0.4$$

Clause 6.3.2.2(4)

Hence, lateral torsional buckling effects may be ignored, and $\chi_{LT} = 1.0$ (*clause 6.3.2.2(4)*).

Lateral torsional buckling resistance: segment BC

$$M_{b,Rd} = \chi_{LT} W_y \frac{f_y}{\gamma_{M1}} \tag{6.55}$$

$$= 1.0 \times 491 \times 10^3 \times (355/1.0)$$

$$= 174.3 \times 10^6 \, \text{N mm} = 174.3 \, \text{kN m}$$

$$\frac{M_{Ed}}{M_{b,Rd}} = \frac{139.2}{174.3} = 0.80$$

$0.80 \leq 1.0$ $\therefore$ acceptable

Member buckling resistance in combined bending and axial compression (clause 6.3.3)

Members subjected to combined bending and axial compression must satisfy both *equations* (6.61) and (6.62).

$$\frac{N_{Ed}}{\chi_y N_{Rk}/\gamma_{M1}} + k_{yy}\frac{M_{y,Ed}}{\chi_{LT}M_{y,Rk}/\gamma_{M1}} + k_{yz}\frac{M_{z,Ed}}{M_{z,Rk}/\gamma_{M1}} \leq 1 \tag{6.61}$$

$$\frac{N_{Ed}}{\chi_z N_{Rk}/\gamma_{M1}} + k_{zy}\frac{M_{y,Ed}}{\chi_{LT}M_{y,Rk}/\gamma_{M1}} + k_{zz}\frac{M_{z,Ed}}{M_{z,Rk}/\gamma_{M1}} \leq 1 \tag{6.62}$$

Determination of interaction factors k_{ij} (Annex A)

For this example, alternative method 1 (*Annex A*) will be used for the determination of the interaction factors k_{ij}. There is no need to consider k_{yz} and k_{zz} in this case, since $M_{z,Ed} = 0$.

For Class 1 and 2 cross-sections

$$k_{yy} = C_{my}C_{mLT}\frac{\mu_y}{1 - N_{Ed}/N_{cr,y}}\frac{1}{C_{yy}}$$

$$k_{zy} = C_{my}C_{mLT}\frac{\mu_y}{1 - N_{Ed}/N_{cr,y}}\frac{1}{C_{zy}}0.6\sqrt{\frac{w_y}{w_z}}$$

Non-dimensional slendernesses

From the flexural buckling check:

$$\bar{\lambda}_y = 1.42 \quad\text{and}\quad \bar{\lambda}_z = 0.84 \quad \therefore \bar{\lambda}_{max} = 1.42$$

From the lateral torsional buckling check:

$$\bar{\lambda}_{LT} = 0.23 \quad\text{and}\quad \bar{\lambda}_0 = 0.23$$

Equivalent uniform moment factors C_{mi}

Torsional deformation is possible ($\bar{\lambda}_0 > 0$).

From the bending moment diagram, $\psi_y = 1.0$.

Therefore, from *Table A.2*,

$$C_{my,0} = 0.79 + 0.21\psi_y + 0.36(\psi_y - 0.33)\frac{N_{Ed}}{N_{cr,y}}$$

$$= 0.79 + (0.21 \times 1.0) + 0.36 \times (1.0 - 0.33)\frac{90}{1470} = 1.01$$

$C_{mz,0} = C_{mz}$ need not be considered since $M_{z,Ed} = 0$.

$$\varepsilon_y = \frac{M_{y,Ed}}{N_{Ed}}\frac{A}{W_{el,y}} \quad\text{for Class 1, 2 and 3 cross-sections}$$

$$= \frac{139.2 \times 10^6}{90 \times 10^3}\frac{8300}{368\,000} = 34.9$$

$$a_{LT} = 1 - \frac{I_T}{I_y} \geq 1.0 = 1 - \frac{29\,820\,000}{36\,780\,000} = 0.189$$

The elastic torsional buckling force (see Section 13.7 of this guide)

$$N_{cr,T} = \frac{1}{i_0^2}\left(GI_T + \frac{\pi^2 EI_w}{l_T^2}\right) \tag{D13.17}$$

$$i_y = (I_y/A)^{0.5} = (36\,780\,000/8300)^{0.5} = 66.6\text{ mm}$$

$i_z = (I_z/A)^{0.5} = (11\,470\,000/8300)^{0.5} = 37.2\,\text{mm}$

$y_0 = z_0 = 0$ (since the shear centre and centroid of gross cross-section coincide)

$i_0^2 = i_y^2 + i_z^2 + y_0^2 + z_0^2 = 66.6^2 + 37.2^2 = 5813\,\text{mm}^2$

Since the section is closed, the warping contribution is negligible and will be ignored.

$$\therefore N_{cr,T} = \frac{1}{5813}(81\,000 \times 29\,820\,000) = 415502 \times 10^3\,\text{N} = 415\,502\,\text{kN}$$

$$C_{my} = C_{my,0} + (1 - C_{my,0})\frac{\sqrt{\varepsilon_y}a_{LT}}{1 + \sqrt{\varepsilon_y}a_{LT}}$$

$$= 1.01 + (1 - 1.01)\frac{\sqrt{34.9} \times 0.189}{1 + (\sqrt{34.9} \times 0.189)} = 1.01$$

$$C_{mLT} = C_{my}^2 \frac{a_{LT}}{\sqrt{[1 - (N_{Ed}/N_{cr,z})][1 - (N_{Ed}/N_{cr,T})]}}$$

$$= 1.01^2 \frac{0.189}{\sqrt{[1 - (90/4127)][1 - (90/415502)]}} \geq 1.0 \quad (\text{but} \geq 1.0) \quad \therefore C_{mLT} = 1.00$$

Other auxiliary terms

Only the auxiliary terms that are required for the determination of k_{yy} and k_{zy} are calculated:

$$\mu_y = \frac{1 - (N_{Ed}/N_{cr,y})}{1 - \chi_y(N_{Ed}/N_{cr,y})} = \frac{1 - (90/1470)}{1 - 0.41 \times (90/1470)} = 0.96$$

$$\mu_z = \frac{1 - (N_{Ed}/N_{cr,z})}{1 - \chi_z(N_{Ed}/N_{cr,z})} = \frac{1 - (90/4127)}{1 - 0.77 \times (90/4127)} = 0.99$$

$$w_y = \frac{W_{pl,y}}{W_{el,y}} \leq 1.5 = \frac{491\,000}{368\,000} = 1.33$$

$$w_z = \frac{W_{pl,z}}{W_{el,z}} \leq 1.5 = \frac{290\,000}{229\,000} = 1.27$$

$$n_{pl} = \frac{N_{Ed}}{N_{Rk}/\gamma_{M1}} = \frac{90}{2946/1.0} = 0.03$$

$$b_{LT} = 0.5a_{LT}\bar{\lambda}_0^2 \frac{M_{y,Ed}}{\chi_{LT}M_{pl,y,Rd}}\frac{M_{z,Ed}}{M_{pl,z,Rd}} = 0 \quad (\text{because } M_{z,Ed} = 0)$$

$$d_{LT} = 2a_{LT}\frac{\bar{\lambda}_0}{0.1 + \bar{\lambda}_z^4}\frac{M_{y,Ed}}{C_{my}\chi_{LT}M_{pl,y,Rd}}, \frac{M_{z,Ed}}{C_{mz}M_{pl,z,Rd}} = 0 \quad (\text{because } M_{z,Ed} = 0)$$

C_{ij} factors

$$C_{yy} = 1 + (w_y - 1)\left[\left(2 - \frac{1.6}{w_y}C_{my}^2\bar{\lambda}_{max} - \frac{1.6}{w_y}C_{my}^2\bar{\lambda}_{max}^2\right)n_{pl} - b_{LT}\right]$$

$$\geq \frac{W_{el,y}}{W_{pl,y}} = 1 + (1.33 - 1)$$

$$\times \left\{\left[\left(2 - \frac{1.6}{1.33} \times 1.01^2 \times 1.42\right) - \left(\frac{1.6}{1.33} \times 1.01^2 \times 1.42^2\right)\right] \times 0.03 - 0\right\}$$

$$= 0.98 \quad \left(\geq \frac{368\,000}{491\,000} = 0.75\right) \quad \therefore C_{yy} = 0.98$$

$$C_{zy} = 1 + (w_y - 1)\left[\left(2 - 14\frac{C_{my}^2 \bar{\lambda}_{max}^2}{w_y^5}\right)n_{pl} - d_{LT}\right] \geq 0.6\sqrt{\frac{w_y}{w_z}}\frac{W_{el,y}}{W_{pl,y}}$$

$$= 1 + (1.33 - 1) \times \left[\left(2 - 14 \times \frac{1.01^2 \times 1.42^2}{1.33^5}\right) \times 0.03 - 0\right]$$

$$= 0.95 \quad \left(\geq 0.6 \times \sqrt{\frac{1.33}{1.27}}\frac{368\,000}{491\,000} = 0.46\right) \quad \therefore C_{zy} = 0.95$$

Interaction factors k_{ij}

$$k_{yy} = C_{my}C_{mLT}\frac{\mu_y}{1 - N_{Ed}/N_{cr,y}}\frac{1}{C_{yy}}$$

$$= 1.01 \times 1.00 \times \frac{0.96}{1 - 90/1470} \times \frac{1}{0.98} = 1.06$$

$$k_{zy} = C_{my}C_{mLT}\frac{\mu_z}{1 - N_{Ed}/N_{cr,y}}\frac{1}{C_{zy}}0.6\sqrt{\frac{w_y}{w_z}}$$

$$= 1.01 \times 1.00 \times \frac{0.99}{1 - 90/1470} \times \frac{1}{0.95} \times 0.6 \times \sqrt{\frac{1.33}{1.27}} = 0.69$$

Check compliance with interaction formulae (*equations (6.61)* and *(6.62)*)

$$\frac{N_{Ed}}{\chi_y N_{Rk}/\gamma_{M1}} + k_{yy}\frac{M_{y,Ed}}{\chi_{LT}M_{y,Rk}/\gamma_{M1}} + k_{yz}\frac{M_{z,Ed}}{M_{z,Rk}/\gamma_{M1}} \leq 1 \qquad (6.61)$$

$$\Rightarrow \frac{90}{(0.41 \times 2947)/1.0} + 1.06 \times \frac{139.2}{(1.0 \times 174.3)/1.0} = 0.07 + 0.84 = 0.92$$

$0.92 \leq 1.0 \quad \therefore$ *equation (6.61)* is satisfied

$$\frac{N_{Fd}}{\chi_z N_{Rk}/\gamma_{M1}} + k_{zy}\frac{M_{y,Ed}}{\chi_{LT}M_{y,Rk}/\gamma_{M1}} + k_{zz}\frac{M_{z,Ed}}{M_{z,Rk}/\gamma_{M1}} \leq 1 \qquad (6.62)$$

$$\Rightarrow \frac{90}{(0.77 \times 2947)/1.0} + 0.69 \times \frac{139.2}{(1.0 \times 174.3)/1.0} = 0.04 + 0.55 = 0.59$$

$0.59 \leq 1.0 \quad \therefore$ *equation (6.62)* is satisfied

Therefore, a hot-rolled $200 \times 100 \times 16$ RHS in grade S355 steel is suitable for this application.

For comparison, from the *Annex B* method,

$$k_{yy} = 1.06 \qquad k_{zy} = 1.00$$

which gives, for *equation (6.61)*,

$$0.07 + 0.85 = 0.92 \qquad (0.92 \leq 1.0 \therefore \text{ acceptable})$$

and, for *equation (6.62)*,

$$0.04 + 0.80 = 0.83 \qquad (0.83 \leq 1.0 \therefore \text{ acceptable})$$

Example 6.10 considers member resistance under combined bi-axial bending and axial load, and uses alternative method 2 (*Annex B*) to determine the necessary interaction factors k_{ij}.

Example 6.10: member resistance under combined bi-axial bending and axial compression

An H-section member of length 4.2 m is to be designed as a ground-floor column in a multi-storey building. The frame is moment resisting in-plane and braced out-of-plane. The column is subjected to major axis bending due to horizontal forces and minor axis bending due to eccentric loading from the floor beams. From the structural analysis, the design action effects of Figure 6.29 arise in the column.

Figure 6.29. Design action effects on an H-section column

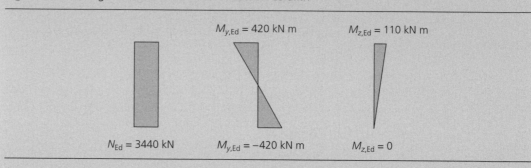

Assess the suitability of a hot-rolled $305 \times 305 \times 240$ H-section in grade S275 steel for this application.

For this example, the interaction factors k_{ij} (for member checks under combined bending and axial compression) will be determined using alternative method 2 (*Annex B*), which is discussed in Chapter 9 of this guide.

Section properties
The section properties are given in Figure 6.30.

Figure 6.30. Section properties for a $305 \times 305 \times 240$ H-section

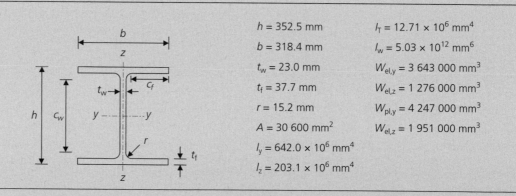

$h = 352.5$ mm $\qquad I_T = 12.71 \times 10^6$ mm^4

$b = 318.4$ mm $\qquad I_w = 5.03 \times 10^{12}$ mm^6

$t_w = 23.0$ mm $\qquad W_{el,y} = 3\,643\,000$ mm^3

$t_f = 37.7$ mm $\qquad W_{el,z} = 1\,276\,000$ mm^3

$r = 15.2$ mm $\qquad W_{pl,y} = 4\,247\,000$ mm^3

$A = 30\,600$ mm^2 $\qquad W_{el,z} = 1\,951\,000$ mm^3

$I_y = 642.0 \times 10^6$ mm^4

$I_z = 203.1 \times 10^6$ mm^4

For a nominal material thickness ($t_f = 37.7$ mm and $t_w = 23.0$ mm) of greater than 16 mm but less than or equal to 40 mm the nominal value of yield strength f_y for grade S275 steel is found from EN 10025-2 to be 265 N/mm^2.

Clause 3.2.6

From *clause 3.2.6*:

$E = 210\,000$ N/mm^2

$G \approx 81\,000$ N/mm^2

Clause 5.5.2

Cross-section classification (clause 5.5.2)

$\varepsilon = \sqrt{235/f_y} = \sqrt{235/265} = 0.94$

Outstand flanges (*Table 5.2*, sheet 2):

$c_f = (b - t_w - 2r)/2 = 132.5$ mm

$c_f/t_f = 132.5/37.7 = 3.51$

Limit for Class 1 flange $= 9\varepsilon = 8.48$

$8.48 > 3.51$ ∴ flanges are Class 1

Web – internal compression part (*Table 5.2*, sheet 1):

$c_w = h - 2t_f - 2r = 246.7$ mm

$c_w/t_w = 246.7/23.0 = 10.73$

Limit for Class 1 web $= 33\varepsilon = 31.08$

$31.08 > 10.73$ ∴ web is Class 1

The overall cross-section classification is therefore Class 1.

Compression resistance of cross-section (clause 6.2.4)

The design compression resistance of the cross-section

$$N_{c,Rd} = \frac{Af_y}{\gamma_{M0}} \qquad \text{for Class 1, 2 or 3 cross-sections} \qquad (6.10)$$

$$= \frac{30\,600 \times 265}{1.00} = 8\,109\,000\,\text{N} = 8109\,\text{kN}$$

8109 kN > 3440 kN   ∴ acceptable

Clause 6.2.4

Bending resistance of cross-section (clause 6.2.5)

Major (y–y) axis
Maximum bending moment

$M_{y,Ed} = 420.0$ kN m

The design major axis bending resistance of the cross-section

$$M_{c,y,Rd} = \frac{W_{pl,y}f_y}{\gamma_{M0}} \qquad \text{for Class 1 or 2 cross-sections} \qquad (6.13)$$

$$= \frac{4\,247\,000 \times 265}{1.00} = 1125 \times 10^6\,\text{N mm} = 1125\,\text{kN m}$$

1125 kN m > 420.0 kN m   ∴ acceptable

Clause 6.2.5

Minor (z–z) axis
Maximum bending moment

$M_{y,Ed} = 110.0$ kN m

The design minor axis bending resistance of the cross-section

$$M_{c,z,Rd} = \frac{W_{pl,z}f_y}{\gamma_{M0}} = \frac{1\,951\,000 \times 265}{1.00} = 517.0 \times 10^6\,\text{N mm} = 517.0\,\text{kN m}$$

517.0 kN m > 110.0 kN m   ∴ acceptable

Shear resistance of cross-section (clause 6.2.6)

The design plastic shear resistance of the cross-section

$$V_{pl,Rd} = \frac{A_v(f_y/\sqrt{3})}{\gamma_{M0}} \qquad (6.18)$$

Clause 6.2.6

Load parallel to web
Maximum shear force

$$V_{Ed} = 840/4.2 = 200 \, kN$$

For a rolled H-section, loaded parallel to the web, the shear area A_v is given by

$$A_v = A - 2bt_f + (t_w + 2r)t_f \quad \text{(but not less than } \eta h_w t_w)$$

Clause NA.2.4 $\eta = 1.0$ from *clause NA.2.4* of the UK National Annex to EN 1993-1-5.

$$h_w = (h - 2t_f) = 352.5 - (2 \times 37.7) = 277.1 \, mm$$

$$\therefore \; A_v = 30\,600 - (2 \times 318.4 \times 37.7) + (23.0 + [2 \times 15.2]) \times 37.7$$

$$= 8606 \, mm^2 \; \text{(but not less than } 1.0 \times 277.1 \times 23.0 = 6373 \, mm^2)$$

$$\therefore \; V_{pl,Rd} = \frac{8606 \times (265/\sqrt{3})}{1.00} = 1317 \times 10^3 \, N = 1317 \, kN$$

$$1317 \, kN > 200 \, kN \qquad \therefore \; \text{acceptable}$$

Load parallel to flanges
Maximum shear force

$$V_{Ed} = 110/4.2 = 26.2 \, kN$$

No guidance on the determination of the shear area for a rolled I- or H-section loaded parallel to the flanges is presented in EN 1993-1-1, although it may be assumed that adopting the recommendations provided for a welded I- or H-section would be acceptable.

The shear area A_v is therefore taken as

$$A_w = A - \sum (h_w t_w) = 30\,600 - (277.1 \times 23.0) = 24\,227 \, mm^2$$

$$\therefore \; V_{pl,Rd} = \frac{24\,227 \times (265/\sqrt{3})}{1.00} = 3707 \times 10^3 \, N = 3707 \, kN$$

$$3707 \, kN > 26.2 \, kN \qquad \therefore \; \text{acceptable}$$

Shear buckling
Shear buckling need not be considered, provided

$$\frac{h_w}{t_w} \leq 72\frac{\varepsilon}{\eta} \qquad \text{for unstiffened webs}$$

Clause NA.2.4 $\eta = 1.0$ from *clause NA.2.4* of the UK National Annex to EN 1993-1-5.

$$72\frac{\varepsilon}{\eta} = 72 \times \frac{0.94}{1.0} = 67.8$$

Actual $h_w/t_w = 277.1/23.0 = 12.0$

$$12.0 \leq 67.8 \qquad \therefore \; \text{no shear buckling check required}$$

Clause 6.2.10 ### Cross-section resistance under bending, shear and axial force (clause 6.2.10)
Provided the shear force V_{Ed} is less than 50% of the design plastic shear resistance $V_{pl,Rd}$ and provided shear buckling is not a concern, then the cross-section need only satisfy the require-
Clause 6.2.9 ments for bending and axial force (*clause 6.2.9*).

In this case, $V_{Ed} < 0.5V_{pl,Rd}$ for both axes, and shear buckling is not a concern (see above). Therefore, the cross-section need only be checked for bending and axial force.

No reduction to the major axis plastic resistance moment due to the effect of axial force is required when both of the following criteria are satisfied:

$$N_{Ed} \leq 0.25N_{pl,Rd} \tag{6.33}$$

$$N_{Ed} \leq \frac{0.5 h_w t_w f_y}{\gamma_{M0}} \qquad (6.34)$$

$$0.25 N_{pl,Rd} = 0.25 \times 8415 = 2104 \, kN$$

$$3440 \, kN > 2104 \, kN \qquad \therefore \text{ equation (6.33) is not satisfied}$$

$$\frac{0.5 h_w t_w f_y}{\gamma_{M0}} = \frac{0.5 \times 277.1 \times 23.0 \times 265}{1.0} = 844.5 \times 10^3 \, N = 844.5 \, kN$$

$$3440 \, kN > 844.5 \, kN \qquad \therefore \text{ equation (6.34) is not satisfied}$$

Therefore, allowance for the effect of axial force on the major axis plastic moment resistance of the cross-section must be made.

No reduction to the minor axis plastic resistance moment due to the effect of axial force is required when the following criterion is satisfied:

$$N_{Ed} \leq \frac{h_w t_w f_y}{\gamma_{M0}} \qquad (6.35)$$

$$\frac{h_w t_w f_y}{\gamma_{M0}} = \frac{277.1 \times 23.0 \times 265}{1.0} = 1689 \times 10^3 \, N = 1689 \, kN$$

$$3440 \, kN > 1689 \, kN \qquad \therefore \text{ equation (6.35) is not satisfied}$$

Therefore, allowance for the effect of axial force on the minor axis plastic moment resistance of the cross-section must be made.

Reduced plastic moment resistances (*clause 6.2.9.1(5)*)
Major (y–y) axis:

Clause 6.2.9.1(5)

$$M_{N,y,Rd} = M_{pl,y,Rd} \frac{1-n}{1-0.5a} \qquad (\text{but } M_{N,y,Rd} \leq M_{pl,y,Rd}) \qquad (6.36)$$

where

$$n = N_{Ed}/N_{pl,Rd} = 3440/8109 = 0.42$$

$$a = (A - 2bt_f)/A = [30\,600 - (2 \times 318.4 \times 37.7)]/30\,600 = 0.22$$

$$\Rightarrow M_{N,y,Rd} = 1125 \times \frac{1 - 0.42}{1 - (0.5 \times 0.22)} = 726.2 \, kN\,m$$

$$726.2 \, kN\,m > 420 \, kN\,m \qquad \therefore \text{ acceptable}$$

Minor (z–z) axis:

For $n > a$

$$M_{N,z,Rd} = M_{pl,z,Rd}\left[1 - \left(\frac{n-a}{1-a}\right)^2\right] \qquad (6.38)$$

$$\Rightarrow M_{N,z,Rd} = 517.0 \times \left[1 - \left(\frac{0.42 - 0.22}{1 - 0.22}\right)^2\right] = 480.4 \, kN\,m$$

$$480.4 \, kN\,m > 110 \, kN\,m \qquad \therefore \text{ acceptable}$$

Cross-section check for bi-axial bending (with reduced moment resistances)

$$\left(\frac{M_{y,Ed}}{M_{N,y,Rd}}\right)^{\alpha} + \left(\frac{M_{z,Ed}}{M_{N,z,Rd}}\right)^{\beta} \leq 1 \tag{6.41}$$

For I- and H-sections:

$\alpha = 2$ and $\beta = 5n$ (but $\beta \geq 1$) = $(5 \times 0.42) = 2.12$

$$\Rightarrow \left(\frac{420}{726.2}\right)^{2} + \left(\frac{110}{480.4}\right)^{2.12} = 0.38$$

$0.38 \leq 1 \qquad \therefore$ acceptable

Clause 6.3.1

Member buckling resistance in compression (clause 6.3.1)

$$N_{b,Rd} = \frac{\chi A f_{y}}{\gamma_{M1}} \qquad \text{for Class 1, 2 and 3 cross-sections} \tag{6.47}$$

$$\chi = \frac{1}{\Phi + \sqrt{\Phi^{2} - \bar{\lambda}^{2}}} \qquad \text{but } \chi \leq 1.0 \tag{6.49}$$

where

$$\Phi = 0.5[1 + \alpha(\bar{\lambda} - 0.2) + \bar{\lambda}^{2}]$$

$$\bar{\lambda} = \sqrt{\frac{A f_{y}}{N_{cr}}} \qquad \text{for Class 1, 2 and 3 cross-sections}$$

Elastic critical force and non-dimensional slenderness for flexural buckling
For buckling about the major (y–y) axis:

$L_{cr} = 0.7L = 0.7 \times 4.2 = 2.94$ m (see Table 6.6)

For buckling about the minor (z–z) axis:

$L_{cr} = 1.0L = 1.0 \times 4.2 = 4.20$ m (see Table 6.6)

$$N_{cr,y} = \frac{\pi^{2} E I_{y}}{L_{cr}^{2}} = \frac{\pi^{2} \times 210\,000 \times 642.0 \times 10^{6}}{2940^{2}} = 153\,943 \times 10^{3}\,\text{N} = 153\,943\,\text{kN}$$

$$\therefore \bar{\lambda}_{y} = \sqrt{\frac{30\,600 \times 265}{153\,943 \times 10^{3}}} = 0.23$$

$$N_{cr,z} = \frac{\pi^{2} E I_{z}}{L_{cr}^{2}} = \frac{\pi^{2} \times 210\,000 \times 203.1 \times 10^{6}}{4200^{2}} = 23\,863 \times 10^{3}\,\text{N} = 23\,863\,\text{kN}$$

$$\therefore \bar{\lambda}_{z} = \sqrt{\frac{30\,600 \times 265}{23\,863 \times 10^{3}}} = 0.58$$

Selection of buckling curve and imperfection factor α
For a hot-rolled H-section (with $h/b \leq 1.2$, $t_{f} \leq 100$ mm and S275 steel):

- for buckling about the y–y axis, use curve b (Table 6.5 (*Table 6.2* of EN 1993-1-1))
- for buckling about the z–z axis, use curve c (Table 6.5 (*Table 6.2* of EN 1993-1-1))
- for curve b, $\alpha = 0.34$ and for curve c, $\alpha = 0.49$ (Table 6.4 (*Table 6.1* of EN 1993-1-1)).

Buckling curves: major (y–y) axis

$$\Phi_y = 0.5 \times [1 + 0.34 \times (0.23 - 0.2) + 0.23^2] = 0.53$$

$$\chi_y = \frac{1}{0.53 + \sqrt{0.53^2 - 0.23^2}} = 0.99$$

$$\therefore N_{b,y,Rd} = \frac{0.99 \times 30\,600 \times 265}{1.0} = 8024 \times 10^3\,\text{N} = 8024\,\text{kN}$$

$8024\,\text{kN} > 3440\,\text{kN}$ $\therefore$ major axis flexural buckling resistance is acceptable

Buckling curves: minor (z–z) axis

$$\Phi_z = 0.5 \times [1 + 0.49 \times (0.58 - 0.2) + 0.58^2] = 0.76$$

$$\chi_z = \frac{1}{0.76 + \sqrt{0.76^2 - 0.58^2}} = 0.80$$

$$\therefore N_{b,z,Rd} = \frac{0.80 \times 30\,600 \times 265}{1.0} = 6450 \times 10^3\,\text{N} = 6450\,\text{kN}$$

$6450\,\text{kN} > 3440\,\text{kN}$ $\therefore$ minor axis flexural buckling resistance is acceptable

Member buckling resistance in bending (clause 6.3.2)

Clause 6.3.2

The 4.2 m column is unsupported along its length with no torsional or lateral restraints. Equal and opposite design end moments of 420 kN m are applied about the major axis. The full length of the column will therefore be checked for lateral torsional buckling.

$$M_{Ed} = 420.0\,\text{kN m}$$

$$M_{b,Rd} = \chi_{LT} W_y \frac{f_y}{\gamma_{M1}} \tag{6.55}$$

where $W_y = W_{pl,y}$ for Class 1 and 2 cross-sections.

Determine M_{cr} ($kL = 4200$ mm)

$$M_{cr} = C_1 \frac{\pi^2 EI_z}{kL^2} \left(\frac{I_w}{I_z} + \frac{kL^2 GI_T}{\pi^2 EI_z} \right)^{0.5} \tag{D6.12}$$

For equal and opposite end moments ($\psi = -1$), $C_1 = 2.752$ (from Table 6.10).

$$\therefore M_{cr} = 2.752 \times \frac{\pi^2 \times 210\,000 \times 203.1 \times 10^6}{4200^2}$$

$$\times \left(\frac{5.03 \times 10^{12}}{203.1 \times 10^6} + \frac{4200^2 \times 81\,000 \times 12.71 \times 10^6}{\pi^2 \times 210\,000 \times 203.1 \times 10^6} \right)^{0.5}$$

$$= 17\,114 \times 10^6\,\text{N mm} = 17\,114\,\text{kN m}$$

Non-dimensional lateral torsional slenderness $\bar{\lambda}_{LT}$: segment BC

$$\bar{\lambda}_{LT} = \sqrt{\frac{W_y f_y}{M_{cr}}} = \sqrt{\frac{4\,247\,000 \times 265}{17\,114 \times 10^6}} = 0.26$$

$$\bar{\lambda}_{LT} \leq \bar{\lambda}_{LT,0} = 0.4$$

Hence, lateral torsional buckling effects may be ignored, and $\chi_{LT} = 1.0$ (*clause 6.3.2.2(4)*).

Clause 6.3.2.2(4)

Lateral torsional buckling resistance

$$M_{b,Rd} = \chi_{LT} W_y \frac{f_y}{\gamma_{M1}} \tag{6.55}$$

$$= 1.0 \times 4\,247\,000 \times (265/1.0)$$

$$= 1125 \times 10^6\,\text{N mm} = 1125\,\text{kN m}$$

$$\frac{M_{Ed}}{M_{b,Rd}} = \frac{420.0}{1125} = 0.37$$

$$0.37 \leq 1.0 \qquad \therefore \text{ acceptable}$$

Clause 6.3.3

Member buckling resistance in combined bending and axial compression (clause 6.3.3)

Members subjected to combined bending and axial compression must satisfy both *equations (6.61)* and *(6.62)*.

$$\frac{N_{Ed}}{\chi_y N_{Rk}/\gamma_{M1}} + k_{yy} \frac{M_{y,Ed}}{\chi_{LT} M_{y,Rk}/\gamma_{M1}} + k_{yz} \frac{M_{z,Ed}}{M_{z,Rk}/\gamma_{M1}} \leq 1 \tag{6.61}$$

$$\frac{N_{Ed}}{\chi_z N_{Rk}/\gamma_{M1}} + k_{zy} \frac{M_{y,Ed}}{\chi_{LT} M_{y,Rk}/\gamma_{M1}} + k_{zz} \frac{M_{z,Ed}}{M_{z,Rk}/\gamma_{M1}} \leq 1 \tag{6.62}$$

Determination of interaction factors k_{ij} (Annex B)

For this example, alternative method 2 (*Annex B*) will be used for the determination of the interaction factors k_{ij}. For axial compression and bi-axial bending, all four interaction coefficients k_{yy}, k_{yz}, k_{zy} and k_{zz} are required.

The column is laterally and torsionally unrestrained, and is therefore susceptible to torsional deformations. Accordingly, the interaction factors should be determined with initial reference to *Table B.2*.

Equivalent uniform moment factors C_{mi} (Table B.3)

Since there is no loading between restraints, all three equivalent uniform moment factors C_{my}, C_{mz} and C_{mLT} may be determined from the expression given in the first row of *Table B.3*, as follows:

$$C_{mi} = 0.6 + 0.4\psi \geq 0.4$$

Considering y–y bending and in-plane supports:

$$\psi = -1, \; C_{my} = 0.6 + (0.4 \times -1) = 0.2 \text{ (but} \geq 0.4) \qquad \therefore \; C_{my} = 0.40$$

Considering z–z bending and in-plane supports:

$$\psi = 0, \; C_{mz} = 0.6 + (0.4 \times 0) = 0.6 \qquad \therefore \; C_{mz} = 0.60$$

Considering y–y bending and out-of-plane supports:

$$\psi = -1, \; C_{mLT} = 0.6 + [0.4 \times (-1)] = 0.2 \text{ (but} \geq 0.4) \qquad \therefore \; C_{mLT} = 0.40$$

Interaction factors k_{ij} (Table B.2 (and Table B.1))

For Class 1 and 2 I sections:

$$k_{yy} = C_{my}\left(1 + (\bar{\lambda}_y - 0.2)\frac{N_{Ed}}{\chi_y N_{Rk}/\gamma_{M1}}\right) \leq C_{my}\left(1 + 0.8\frac{N_{Ed}}{\chi_y N_{Rk}/\gamma_{M1}}\right)$$

$$= 0.40 \times \left(1 + (0.23 - 0.2)\frac{3440}{(0.99 \times 8109)/1.0}\right) = 0.41$$

$$\leq 0.40 \times \left(1 + 0.8\frac{3440}{0.99 \times 8109/1.0}\right) = 0.54 \qquad \therefore \; k_{yy} = 0.41$$

$$k_{zz} = C_{mz}\left(1 + (2\bar{\lambda}_z - 0.6)\frac{N_{Ed}}{\chi_z N_{Rk}/\gamma_{M1}}\right) \leq C_{my}\left(1 + 1.4\frac{N_{Ed}}{\chi_z N_{Rk}/\gamma_{M1}}\right)$$

$$= 0.60 \times \left(1 + [(2 \times 0.58) - 0.6]\frac{3440}{(0.80 \times 8109)/1.0}\right) = 0.78$$

$$\leq 0.60 \times \left(1 + 1.4\frac{3440}{0.79 \times 8415/1.0}\right) = 1.04 \qquad \therefore k_{zz} = 0.78$$

$$k_{yz} = 0.6k_{zz} = 0.6 \times 0.72 = 0.47 \qquad \therefore k_{yz} = 0.47$$

$$k_{zy} = 1 - \frac{0.1\bar{\lambda}_z}{C_{mLT} - 0.25}\frac{N_{Ed}}{\chi_z N_{Rk}/\gamma_{M1}}$$

$$\geq 1 - \frac{0.1}{C_{mLT} - 0.25}\frac{N_{Ed}}{\chi_z N_{Rk}/\gamma_{M1}} \qquad \text{for } \bar{\lambda}_z \geq 0.4$$

$$= 1 - \frac{0.1 \times 0.59}{0.40 - 0.25}\frac{3440}{(0.80 \times 8109)/1.0} = 0.79$$

$$\geq 1 - \frac{0.1}{0.40 - 0.25}\frac{3440}{(0.80 \times 8109)/1.0} = 0.64 \qquad \therefore k_{zy} = 0.79$$

Check compliance with interaction formulae (*equations (6.61)* and *(6.62)*)

$$\frac{N_{Ed}}{\chi_y N_{Rk}/\gamma_{M1}} + k_{yy}\frac{M_{y,Ed}}{\chi_{LT} M_{y,Rk}/\gamma_{M1}} + k_{yz}\frac{M_{z,Ed}}{M_{z,Rk}/\gamma_{M1}} \leq 1 \qquad (6.61)$$

$$\Rightarrow \frac{3440}{(0.99 \times 8109)/1.0} + 0.41 \times \frac{420.0}{(1.0 \times 1125)/1.0} + 0.47 \times \frac{110.0}{517.0/1.0} = 0.43 + 0.15 + 0.10$$

$$= 0.68$$

$$0.68 \leq 1.0 \qquad \therefore \textit{equation (6.61)} \text{ is satisfied}$$

$$\frac{N_{Ed}}{\chi_z N_{Rk}/\gamma_{M1}} + k_{zy}\frac{M_{y,Ed}}{\chi_{LT} M_{y,Rk}/\gamma_{M1}} + k_{zz}\frac{M_{z,Ed}}{M_{z,Rk}/\gamma_{M1}} \leq 1 \qquad (6.62)$$

$$\Rightarrow \frac{3440}{(0.80 \times 8109)/1.0} + 0.79 \times \frac{420.0}{(1.0 \times 1125)/1.0} + 0.78 \times \frac{110.0}{517.0/1.0} = 0.53 + 0.30 + 0.17$$

$$= 1.0$$

$$1.0 \leq 1.0 \qquad \therefore \textit{equation (6.62)} \text{ is satisfied}$$

Therefore, a hot-rolled $305 \times 305 \times 240$ H-section in grade S275 steel is suitable for this application.

For comparison, from the *Annex A* method:

$$k_{yy} = 0.74 \qquad k_{yz} = 0.49 \qquad k_{zy} = 0.43 \qquad k_{zz} = 1.33$$

which gives, for *equation (6.61)*,

$$0.43 + 0.28 + 0.10 = 0.81 \qquad (0.81 \leq 1.0 \; \therefore \text{ acceptable})$$

and, for *equation (6.62)*,

$$0.53 + 0.16 + 0.15 = 0.85 \qquad (0.85 \leq 1.0 \; \therefore \text{ acceptable})$$

Columns in simple construction

For columns in frames designed according to the principles of simple construction, i.e. assuming simply supported beams and column moments at each floor level due solely to notional eccentricities of beam reactions, Brettle and Brown (2009) consolidate *equations* (*6.61*) and (*6.62*) into the single expression

$$\frac{N_{Ed}}{N_{b,z,Rd}} + \frac{M_{y,Ed}}{M_{b,Rd}} + 1.5\frac{M_{z,Ed}}{M_{c,z,Rd}} \leq 1.0 \qquad (D6.20)$$

This simplified expression is based on the fact that the axial term is dominant, since the two moment terms ($M_{y,Ed}$ and $M_{z,Ed}$) will be small, and that failure will be about the minor axis. A full explanation of the basis for this approach and limitations on its use is available in NCCI SN048 (SCI, 2006).

6.3.4 General method for lateral and lateral torsional buckling of structural components

Clause 6.3.4

Clause 6.3.4 provides a general method to assess the lateral and lateral torsional buckling resistance of structural components. The method is relatively new, and, as such, has not yet been subjected to the same level and breadth of scrutiny as the more established methods. For the purposes of this guide, it is recommended that the provisions of *clause 6.3.4* are adopted

Clause 6.3.4

Clause NA.2.22

with caution and preferably verified with independent checks. The UK National Annex (*clause NA.2.22*) limits application of the method to straight members subject to in-plane mono-axial bending and/or compression, and states that the global buckling reduction factor χ_{op} should be taken as the minimum value of the buckling reduction for compression χ, and that for bending as χ_{LT}.

6.3.5 Lateral torsional buckling of members with plastic hinges

Two specific requirements for addressing lateral torsional buckling effects in frames designed according to a plastic hinge analysis are listed:

- restraint at plastic hinges
- stable lengths for segments between plastic hinges.

Since the design objective is now to ensure that load carrying of the frame is controlled by the formation of a plastic collapse mechanism, any premature failure due to lateral instability must be prevented. This may be achieved by providing a suitable system of restraints – lateral and/or torsional.

Clause 6.3.5.2

Clause 6.3.5.2 states where restraints are required and the performance necessary from each of them. The rules are very similar to the equivalent provision of BS 5950: Part 1.

A simple check for stable length of member with end moments M and ψM (and negligible axial load) is provided by *equation* (*6.68*) as

$$L_{stable} \not> 35\varepsilon i_z \qquad \text{for } 0.625 \leq \psi \leq 1$$

$$\qquad\qquad\qquad\qquad\qquad\qquad\qquad\qquad (6.68)$$

$$L_{stable} \not> (60 - 40\psi)\varepsilon i_z \qquad \text{for } -1 \leq \psi \leq 0.625$$

Clause BB.3

More detailed rules covering tapered haunches (with two or three flanges) are provided in *clause BB.3*, and are discussed in Chapter 11 of this guide.

6.4. Uniform built-up compression members

Clause 6.4

Clause 6.4 covers the design of uniform built-up compression members. The principal difference between the design of built-up columns and the design of conventional (solid) columns is in their response to shear. In conventional column buckling theory, lateral deflections are assessed (with a suitable level of accuracy) on the basis of the flexural properties of the member, and the effects of shear on deflections are ignored. For built-up columns, shear deformations are far more significant (due to the absence of a solid web), and therefore have to be evaluated and accounted for in the development of design procedures.

There are two distinct types of built-up member (laced and battened), characterised by the layout of the web elements, as shown in Figure 6.31. Laced columns contain diagonal web elements with

Figure 6.31. Types of built-up compression member. (a) Laced column. (b) Battened column

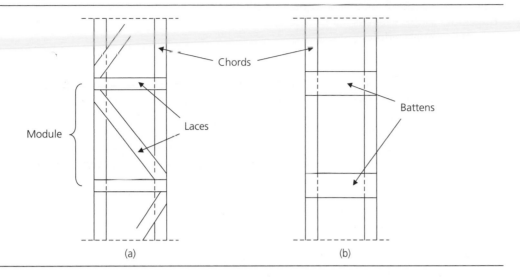

or without additional horizontal web elements; these web elements are generally assumed to have pinned end conditions and therefore to act in axial tension or compression. Battened columns (see Figure 6.32) contain horizontal web elements only and behave in the same manner as Vierendeel trusses, with the battens acting in flexure. Battened struts are generally more flexible in shear than laced struts.

Clause 6.4 also provides guidance for closely spaced built-up members such as back-to-back channels. Background to the analysis and design of built-up structures has been reported by Galambos (1998) and Narayanan (1982).

In terms of material consumption, built-up members can offer much greater efficiency than single members. However, with the added expenses of the fabrication process, and the rather unfashionable aesthetics (often containing corrosion traps), the use of built-up members is less popular nowadays in the UK than in the past. Consequently, BS 5950: Part 1 offers less

Clause 6.4

Figure 6.32. Battened columns

detailed guidance on the subject than Eurocode 3. The basis of the BS 5950 method is also different from the Eurocode approach, with BS 5950 using a modified Euler buckling theory (Engessor, 1909), whereas the Eurocode opts for a second-order analysis with a specified initial geometric imperfection.

6.4.1   General

Clause 6.4

Designing built-up members based on calculations of the discontinuous structure is considered too time-consuming for practical design purposes. *Clause 6.4* offers a simplified model that may be applied to uniform built-up compression members with pinned end conditions (although the code notes that appropriate modifications may be made for other end conditions). Essentially the model replaces the discrete (discontinuous) elements of the built-up column with an equivalent continuous (solid) column, by 'smearing' the properties of the web members (lacings or battens). Design then comprises two steps:

1. Analyse the full 'equivalent' member (with smeared shear stiffness) using second-order theory, as described in the following subsection, to determine maximum design forces and moments.
2. Check critical chord and web members under design forces and moments. Joints must also be checked – see Chapter 12 of this guide.

Clause 6.4.1

The following rules regarding the application of the model are set out in *clause 6.4.1*:

1. The chord members must be parallel.
2. The lacings or battens must form equal modules (i.e. uniform-sized lacings or battens and regular spacing).
3. The minimum number of modules in a member is three.
4. The method is applicable to built-up members with lacings in one or two directions, but is only recommended for members battened in one direction.
5. The chord members may be solid members or themselves built-up (with lacings or battens in the perpendicular plane).

Clause 6.4.1(6)

For global structural analysis purposes a member (bow) imperfection of magnitude $e_0 = L/500$ may be adopted. This magnitude of imperfection is also employed in the design formulations of *clause 6.4.1(6)*, and has an empirical basis.

Design forces in chords and web members

Clause 6.4.1(6)
Clause 6.4.1(7)

Evaluation of the design forces to apply to chord and web members is covered in *clauses 6.4.1(6)* and *6.4.1(7)*, respectively. The maximum design chord forces $N_{ch,Ed}$ are determined from the applied compression forces N_{Ed} and applied bending moments M_{Ed}^I. The formulations were derived from the governing differential equation of a column and by considering second-order effects, resulting in the occurrence of the maximum design chord force at the mid-length of the column.

For a member with two identical chords the design force $N_{ch,Ed}$ should be determined from

$$N_{ch,Ed} = 0.5 N_{Ed} \frac{M_{Ed} h_0 A_{ch}}{2 I_{eff}} \tag{6.69}$$

where

$$M_{Ed} = \frac{N_{Ed} e_0 + M_{Ed}^I}{1 - N_{Ed}/N_{cr} - N_{Ed}/S_v}$$

$$N_{cr} = \frac{\pi^2 E I_{eff}}{L^2} \quad \text{is the effective critical force of the built-up member}$$

N_{Ed} is the design value of the applied compression for on the built-up member
M_{Ed} is the design value of the maximum moment at the mid-length of the built-up member including second-order effects

$M_{\mathrm{Ed}}^{\mathrm{I}}$ is the design value of the applied moment at the mid-length of the built-up member (without second-order effects)

h_0 is the distance between the centroids of the chords

A_{ch} is the cross-sectional area of one chord

I_{eff} is the effective second moment of area of the built-up member (see the following sections)

S_{v} is the shear stiffness of the lacings or battened panel (see the following sections)

e_0 is the assumed imperfection magnitude and may be taken as $L/500$.

It should be noted that although the formulations include an allowance for applied moments $M_{\mathrm{Ed}}^{\mathrm{I}}$, these are intended to cover small incidental bending moments, such as those arising from load eccentricities.

The lacings and battens should be checked at the end panels of the built-up member, where the maximum shear forces occur. The design shear force V_{Ed} should be taken as

$$V_{\mathrm{Ed}} = \pi \frac{M_{\mathrm{Ed}}}{L} \qquad (6.70)$$

where M_{Ed} has been defined above.

6.4.2   Laced compression members

The chords and diagonal lacings of a built-up laced compression member should be checked for buckling in accordance with *clause 6.3.1*. Various recommendations on construction details for laced members are provided in *clause 6.4.2.2*.

Clause 6.3.1

Clause 6.4.2.2

Chords

The design compression force $N_{\mathrm{ch,Ed}}$ in the chords is determined as described in the previous section. This should be shown to be less than the buckling resistance of the chords, based on a buckling length measured between the points of connection of the lacing system.

For lacings in one direction only, the buckling length of the chord L_{ch} may generally be taken as the system length (although reference should be made to *Annex BB*). For lacings in two directions, buckling lengths are defined in the three-dimensional illustrations of *Figure 6.8* of EN 1993-1-1.

Lacings

The design compression force in the lacings may be easily determined from the design shear force V_{Ed} (described in the previous section) by joint equilibrium. Again, this design compressive force should be shown to be less than the buckling resistance. In general, the buckling length of the lacing may be taken as the system length (though, as for chords, reference should be made to *Annex BB*).

Shear stiffness and effective second moment of area

The shear stiffness and effective second moment of area of the lacings required for the determination of the design forces in the chords and lacings are defined in *clauses 6.4.2.1(3)* and *6.4.2.1(4)*.

Clause 6.4.2.1(3)

Clause 6.4.2.1(4)

The shear stiffness S_{v} of the lacings depends upon the lacing layout, and, for the three common arrangements, reference should be made to *Figure 6.9* of EN 1993-1-1.

For laced built-up members, the effective second moment of area may be taken as

$$I_{\mathrm{eff}} = 0.5 h_0^2 A_{\mathrm{ch}} \qquad (6.72)$$

6.4.3   Battened compression members

The chords, battens and joints of battened compression members should be checked under the design forces and moments at mid-length and in an end panel. Various recommendations on design details for battened members are provided in *clause 6.4.3.2*.

Clause 6.4.3.2

Clause 6.4.3.1(2)

The shear stiffness S_v of a battened built-up member is given in *clause 6.4.3.1(2)*, and should be taken as

$$S_v = \frac{24EI_{ch}}{a^2[1 + (2I_{ch}/nI_b)(h_0/a)]} \quad \text{but} \quad \leq \frac{2\pi^2 EI_{ch}}{a^2} \tag{6.73}$$

where

I_{ch} is the in-plane second moment of area of one chord (about its own neutral axis)
I_b is the in-plane second moment of area of one batten (about its own neutral axis).

Clause 6.4.3.1(3)

The effective second moment of area I_{eff} of a battened built-up member is given in *clause 6.4.3.1(3)*, and may be taken as

$$I_{eff} = 0.5h_0^2 A_{ch} + 2\mu I_{ch} \tag{6.74}$$

where μ is a so-called efficiency factor, taken from *Table 6.8* of EN 1993-1-1. The second part of the right-hand side of *equation (6.74)*, $2\mu I_{ch}$, represents the contribution of the moments of inertia of the chords to the overall bending stiffness of the battened member. This contribution is not included for laced columns (see *equation (6.72)*); the primary reason behind this is that the spacing of the chords in battened built-up members is generally rather less than that for laced members, and it can therefore become uneconomical to neglect the chord contribution.

The efficiency factor μ, the value of which may range between zero and unity, controls the level of chord contribution that may be exploited. The recommendations of *Table 6.8* of EN 1993-1-1 were made to ensure 'safe side' theoretical predictions of a series of experimental results (Narayanan, 1982).

6.4.4 Closely spaced built-up members

Clause 6.4.4

Clause 6.4.4 covers the design of closely spaced built-up members. Essentially, provided the chords of the built-up members are either in direct contact with one another or closely spaced and connected through packing plates and the conditions of *Table 6.9* of EN 1993-1-1 are met, the built-up members may be designed as integral members (ignoring shear deformations) following the provisions of *clause 6.3*; otherwise the provisions of the earlier parts of *clause 6.4* apply.

Clause 6.3
Clause 6.4

REFERENCES

Brettle ME and Brown DG (2009) *Steel Building Design: Concise Eurocodes.* Steel Construction Institute, Ascot, P362.

Chen WF and Atsuta T (1977) *Theory of Beam Columns.* McGraw-Hill, New York.

ECCS (1990) *Background Documentation to Eurocode 3: Part 1.1.* European Convention for Constructional Steelwork, Brussels.

Engessor F (1909) Über die Knickfestigkeit von Rahmenstäben. *Zentralblatt der Bauverwaltung*, **29**: 136 [in German].

Galambos TV (ed.) (1998) *Guide to Stability Design Criteria for Metal Structures*, 5th edn. Wiley, New York.

Gardner L (2011) *Steel Building Design: Stability of Beams and Columns.* Steel Construction Institute, Ascot, P360.

Narayanan R (ed.) (1982) *Axially Compressed Structures – Stability and Strength.* Applied Science, Amsterdam.

SCI (2005a) NCCI SN002: Determination of non-dimensional slenderness of I- and H-sections. http://www.steel-ncci.co.uk.

SCI (2005b) NCCI SN003: Elastic critical moment for lateral torsional buckling. http://www.steel-ncci.co.uk.

SCI (2005c) NCCI SN009. Effective lengths and destabilizing load parameters for beams and cantilevers – common cases. http://www.steel-ncci.co.uk.

SCI (2006) NCCI SN048: Verification of columns in simple construction – a simplified interaction criterion. http://www.steel-ncci.co.uk.

SCI/BCSA (2009) *Steel Building Design: Design Data.* Steel Construction Institute, Ascot, P363.

Timoshenko SP and Gere JM (1961) *Theory of Elastic Stability*, 2nd edn. McGraw-Hill, New York.

Trahair NS (1993) *Flexural–torsional Buckling of Structures*. Chapman and Hall, London.

Trahair NS, Bradford MA, Nethercot DA and Gardner L (2008) *The Behaviour and Design of Steel Structures to EC3*, 4th edn. Spon, London.

Designers' Guide to Eurocode 3: Design of Steel Buildings, 2nd ed.
ISBN 978-0-7277-4172-1

ICE Publishing: All rights reserved
doi: 10.1680/dsb.41721.101

Chapter 7
Serviceability limit states

This chapter concerns the subject of serviceability limit states. The material in this chapter is covered in *Section 7* of Eurocode 3 *Part 1.1*, and the following clauses are addressed:

- General   *Clause 7.1*
- Serviceability limit states for buildings   *Clause 7.2*

Overall, the coverage of serviceability considerations in EN 1993-1-1 is very limited, with little explicit guidance provided. However, as detailed below, for further information reference should be made to EN 1990, on the basis that many serviceability criteria are independent of the structural material. For serviceability issues that are material-specific, reference should be made to EN 1992 to EN 1999, as appropriate. Clauses 3.4, 6.5 and A1.4 of EN 1990 contain guidance relevant to serviceability; clause A1.4 of EN 1990 (as with the remainder of Annex A1 of EN 1990) is specific to buildings.

7.1. General

Serviceability limit states are defined in Clause 3.4 of EN 1990 as those that concern:

- the functionality of the structure or structural members under normal use
- the comfort of the people
- the appearance of the structure.

For buildings, the primary concerns are horizontal and vertical deflections and vibrations.

According to clause 3.4 of EN 1990, a distinction should be made between reversible and irreversible serviceability limit states. Reversible serviceability limit states are those that would be infringed on a non-permanent basis, such as excessive vibration or high elastic deflections under temporary (variable) loading. Irreversible serviceability limit states are those that would remain infringed even when the cause of infringement was removed (e.g. permanent local damage or deformations).

Further, three categories of combinations of loads (actions) are specified in EN 1990 for serviceability checks: characteristic, frequent and quasi-permanent. These are given by equations (6.14) to (6.16) of EN 1990, and summarised in Table 7.1 (Table A1.4 of EN 1990), where each combination contains a permanent action component (favourable or unfavourable), a leading variable component and other variable components. Where a permanent action is unfavourable, which is generally the case, the upper characteristic value of a permanent action $G_{\mathrm{kj,sup}}$ should be used; where an action is favourable (such as a permanent action reducing uplift due to wind loading), the lower characteristic value of a permanent action $G_{\mathrm{kj,inf}}$ should be used.

Unless otherwise stated, for all combinations of actions in a serviceability limit state the partial factors should be taken as unity (i.e. the loading should be unfactored). An introduction to EN 1990 is contained in Chapter 14 of this guide, where combinations of actions are discussed in more detail.

The characteristic combination of actions would generally be used when considering the function of the structure and damage to structural and non-structural elements; the frequent combination would be applied when considering the comfort of the user, the functioning of machinery and

Table 7.1. Design values of actions for use in the combination of actions (Table A1.4 of EN 1990)

Combination	Permanent action G_d		Variable actions Q_d	
	Unfavourable	Favourable	Leading	Others
Characteristic	$G_{kj,sup}$	$G_{kj,inf}$	$Q_{k,1}$	$\psi_{0,i}Q_{k,i}$
Frequent	$G_{kj,sup}$	$G_{kj,inf}$	$\psi_{1,1}Q_{k,1}$	$\psi_{2,i}Q_{k,i}$
Quasi-permanent	$G_{kj,sup}$	$G_{kj,inf}$	$\psi_{2,1}Q_{k,1}$	$\psi_{2,i}Q_{k,i}$

avoiding the possibility of ponding of water; the quasi-permanent combination would be used when considering the appearance of the structure and long-term effects (e.g. creep).

The purpose of the ψ factors (ψ_0, ψ_1 and ψ_2) that appear in the load combinations of Table 7.1 is to modify characteristic values of variable actions to give representative values for different situations. Numerical values of the ψ factors are given in Table 14.1 of this guide. Further discussion of the ψ factors may also be found in Chapter 14 of this guide and in *Corrosion Protection of Steel Bridges* (Corus, 2002).

7.2. Serviceability limit states for buildings

It is emphasised in both EN 1993-1-1 and EN 1990 that serviceability limits (e.g. for deflections and vibrations) should be specified for each project and agreed with the client. Numerical values for these limits are not provided in either document, but recommended values are given in *clause NA.2.23* (vertical deflection limits) and *clause NA.2.24* (horizontal deflection limits) of the UK National Annex, as detailed in the following subsections.

Clause NA.2.23
Clause NA.2.24

7.2.1 Vertical deflections

Total vertical deflections w_{tot} are defined in EN 1990 by a number of components (w_c, w_1, w_2 and w_3), as shown in Figure 7.1 (Figure A1.1 of EN 1990), where

w_c	is the precamber in the unloaded structural member
w_1	initial part of the deflection under permanent loads
w_2	long-term part of the deflection under permanent loads
w_3	additional part of the deflection due to variable loads
w_{tot}	total deflection ($w_1 + w_2 + w_3$)
w_{max}	remaining total deflection taking into account the precamber ($w_{tot} - w_c$).

Clause NA.2.23

Clause NA.2.23 of the UK National Annex provides recommended vertical deflection limits (Table 7.2) for serviceability verifications under the characteristic load combination, and it is stated that deflections should be calculated under variable action only (i.e. without permanent actions). In the characteristic load combination (equation (6.14b) of EN 1990), the leading variable action, which will typically be the imposed load when considering vertical deflections, is unfactored – i.e. deflection checks will be typically carried out under unfactored imposed loads. This is generally in line with existing UK practice, although it is also noted in *clause NA.2.23* that there may be circumstances where greater or lesser values for deflection limits will be appropriate. Likewise, there may be applications where designers opt to limit total deflections (i.e. under permanent plus variable actions).

Clause NA.2.23

Figure 7.1. Definitions of vertical deflections

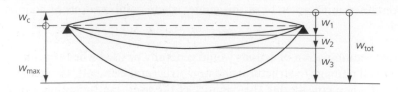

Table 7.2. Vertical deflection limits (from **clause NA 2.23** of the UK National Annex)   *Clause NA 2.23*

Design situation	Deflection limit
Cantilevers	Length/180
Beams carrying plaster or other brittle finish	Span/360
Other beams (except purlins and sheeting rails)	Span/200
Purlins and sheeting rails	To suit cladding

Example 7.1: vertical deflection of beams

A simply supported floor beam in a building of span 5.6 m is subjected to the following (unfactored) loading:

- permanent action: 8.6 kN/m
- variable action (imposed floor action): 32.5 kN/m.

Choose a suitable UB such that the vertical deflection limits of Table 7.2 are not exceeded.

From **clause 3.2.6**

$$E = 210\,000 \text{ N/mm}^2$$

Following **clause NA.2.23** of the UK National Annex, deflections will be checked under unfactored variable actions only.

$$\therefore q = 32.5 \text{ kN/m}$$

Under a uniformly distributed load q, the maximum deflection w of a simply supported beam is

$$w = \frac{5}{384} \frac{qL^4}{EI}$$

$$\Rightarrow I_{required} = \frac{5}{384} \frac{qL^4}{Ew}$$

Selecting a deflection limit of span/200 from Table 7.2:

$$\Rightarrow I_{required} = \frac{5}{384} \frac{qL^4}{Ew} = \frac{5}{384} \times \frac{32.5 \times 5600^4}{210\,000 \times (5600/200)} = 70.8 \times 10^6 \text{ mm}^4$$

From section tables, UKB $356 \times 127 \times 33$ has a second moment of area (about the major axis) I_y of $82.49 \times 10^6 \text{ mm}^4$:

$$82.49 \times 10^6 > 70.8 \times 10^6 \qquad \therefore \text{ UKB } 356 \times 127 \times 33 \text{ is acceptable}$$

If the deflections under the total unfactored load (permanent and variable actions) were to be limited to span/200, the required second moment of area would increase to $89.5 \times 10^6 \text{ mm}^4$.

Clause 3.2.6

Clause NA.2.23

7.2.2 Horizontal deflections

Similar to the treatment of vertical deflections, **clause NA.2.24** of the UK National Annex recommends that horizontal deflections are checked using the characteristic load combination – i.e. under unfactored variable loads. Again, this is in line with existing UK practice. Recommended deflection limits are given in Table 7.3, although, as with the vertical deflection limits, it is noted that greater or lesser values may be appropriate in certain circumstances.

Clause NA.2.24

Figure 7.2. Definitions of horizontal deflections

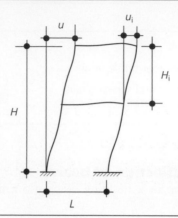

Clause NA 2.23

Table 7.3. Horizontal deflection limits (from *clause NA 2.24* of the UK National Annex)

Design situation	Deflection limit
Tops of columns in single storey buildings, except portal frames	Height/300
Columns in portal frame buildings, not supporting crane runways	To suit cladding
In each storey of a building with more than one storey	Height of storey/300

The EN 1990 notation to describe horizontal deflections is illustrated in Figure 7.2, where u is the total horizontal deflection of a structure of height H, and u_i is the horizontal deflection in each storey (i) of height H_i.

7.2.3 Dynamic effects

Dynamic effects need to be considered in structures to ensure that vibrations do not impair the comfort of the user or the functioning of the structure or structural members. Essentially, this is achieved provided the natural frequencies of vibration are kept above appropriate levels, which depend upon the function of the structure and the source of vibration. Possible sources of vibration include walking, synchronised movements of people, ground-borne vibrations from traffic, and wind action. Further guidance on dynamic effects may be found in EN 1990, *Corrosion Protection of Steel Bridges* (Corus, 2002) and other specialised literature (e.g. Wyatt, 1989).

REFERENCES

Corus (2002) *Corrosion Protection of Steel Bridges*. Corus Construction Centre, Scunthorpe.
Wyatt TA (1989) *Design Guide on the Vibration of Floors*. Steel Construction Institute, Ascot, P076.

Designers' Guide to Eurocode 3: Design of Steel Buildings, 2nd ed.
ISBN 978-0-7277-4172-1

ICE Publishing: All rights reserved
doi: 10.1680/dsb.41721.105

Chapter 8

Annex A (informative) – Method 1: interaction factors k_{ij} for interaction formula in *clause 6.3.3(4)*

For uniform members subjected to combined bending and axial compression, *clause 6.3.3(4)* provides two interaction formulae, both of which must be satisfied. Each of the interaction formulae contains two interaction factors: k_{yy} and k_{yz} for *equation (6.61)* and k_{zy} and k_{zz} for *equation (6.62)*.

Clause 6.3.3(4)

Two alternative methods to determine these four interaction factors (k_{yy}, k_{yz}, k_{zy} and k_{zz}) are given by EN 1993-1-1; Method 1 is contained within *Annex A*, and is described in this chapter, and Method 2 is contained within *Annex B*, and described in Chapter 9 of this guide.

The UK National Annex (*clause NA.3*) allows the use of either *Annex A* or *Annex B*, but limits the application of Annex A to doubly symmetric sections. Of the two methods, Method 1 generally requires more calculation effort, due to the large number of auxiliary terms, while Method 2 is more straightforward. However, Method 1 will generally offer more competitive solutions.

Clause NA.3

Method 1 is based on second-order in-plane elastic stability theory, and maintains consistency with the theory, as far as possible, in deriving the interaction factors. Development of the method has involved an extensive numerical modelling programme. Emphasis has been placed on achieving generality as well as consistency with the individual member checks and cross-section verifications. Inelastic behaviour has been allowed for when considering Class 1 and 2 cross-sections by incorporating plasticity factors that relate the elastic and plastic section moduli. (Further details of the method, developed at the Universities of Liege and Clermont-Ferrand, have been reported by Boissonnade *et al.*, 2002).

The basic formulations for determining the interaction factors using Method 1 are given in Table 8.1 (*Table A.1* of EN 1993-1-1), along with the extensive set of auxiliary terms. The equivalent uniform moment factors $C_{mi,0}$ that depend on the shape of the applied bending moment diagram about each axis together with the support and out-of-plane restraint conditions, are given in Table 8.2 (*Table A.2* of EN 1993-1-1). A distinction is made between members susceptible or not susceptible to lateral–torsional buckling in calculating the factors C_{my}, C_{mz} (both of which represent in-plane behaviour) and C_{mLT} (which represents out-of-plane behaviour).

Method 1 is applied in Example 6.9 to assess the resistance of a rectangular hollow section member under combined axial load and major axis bending.

Table 8.1. Interaction factors k_{ij} for interaction formula in *clause 6.3.3(4)* (*Table A.1* of EN 1993-1-1)

Interaction factors	Design assumptions	
	elastic cross-sectional properties class 3, class 4	plastic cross-sectional properties class 1, class 2
k_{yy}	$C_{my}C_{mLT}\dfrac{\mu_y}{1-\dfrac{N_{Ed}}{N_{cr,y}}}$	$C_{my}C_{mLT}\dfrac{\mu_y}{1-\dfrac{N_{Ed}}{N_{cr,y}}}\dfrac{1}{C_{yy}}$
k_{yz}	$C_{mz}\dfrac{\mu_y}{1-\dfrac{N_{Ed}}{N_{cr,z}}}$	$C_{mz}\dfrac{\mu_y}{1-\dfrac{N_{Ed}}{N_{cr,z}}}\dfrac{1}{C_{yz}}0.6\sqrt{\dfrac{w_z}{w_y}}$
k_{zy}	$C_{my}C_{mLT}\dfrac{\mu_z}{1-\dfrac{N_{Ed}}{N_{cr,y}}}$	$C_{my}C_{mLT}\dfrac{\mu_z}{1-\dfrac{N_{Ed}}{N_{cr,y}}}\dfrac{1}{C_{zy}}0.6\sqrt{\dfrac{w_y}{w_z}}$
k_{zz}	$C_{mz}\dfrac{\mu_z}{1-\dfrac{N_{Ed}}{N_{cr,z}}}$	$C_{mz}\dfrac{\mu_z}{1-\dfrac{N_{Ed}}{N_{cr,z}}}\dfrac{1}{C_{zz}}$

Auxiliary terms:

$$\mu_y = \frac{1-\dfrac{N_{Ed}}{N_{cr,y}}}{1-\chi_y\dfrac{N_{Ed}}{N_{cr,y}}}$$

$$\mu_z = \frac{1-\dfrac{N_{Ed}}{N_{cr,z}}}{1-\chi_z\dfrac{N_{Ed}}{N_{cr,z}}}$$

$$w_y = \frac{W_{pl,y}}{W_{el,y}} \leq 1.5$$

$$w_z = \frac{W_{pl,z}}{W_{el,z}} \leq 1.5$$

$$n_{pl} = \frac{N_{Ed}}{N_{Rk}/\gamma_{M1}}$$

C_{my} see Table A.2

$$a_{LT} = 1 - \frac{I_T}{I_y} \geq 0$$

$$C_{yy} = 1 + (w_y - 1)\left[\left(2 - \frac{1.6}{w_y}C_{my}^2\bar{\lambda}_{max} - \frac{1.6}{w_y}C_{my}^2\bar{\lambda}_{max}^2\right)n_{pl} - b_{LT}\right] \geq \frac{W_{el,y}}{W_{pl,y}}$$

$$\text{with } b_{LT} = 0.5a_{LT}\bar{\lambda}_0^2\frac{M_{y,Ed}}{\chi_{LT}M_{pl,y,Rd}}\frac{M_{z,Ed}}{M_{pl,z,Rd}}$$

$$C_{yz} = 1 + (w_z - 1)\left[\left(2 - 14\frac{C_{mz}^2\bar{\lambda}_{max}^2}{w_z^5}\right)n_{pl} - c_{LT}\right] \geq 0.6\sqrt{\frac{w_z}{w_y}}\frac{W_{el,z}}{W_{pl,z}}$$

$$\text{with } c_{LT} = 10a_{LT}\frac{\bar{\lambda}_0^2}{5+\bar{\lambda}_z^4}\frac{M_{y,Ed}}{C_{my}\chi_{LT}M_{pl,y,Rd}}$$

$$C_{zy} = 1 + (w_y - 1)\left[\left(2 - 14\frac{C_{my}^2\bar{\lambda}_{max}^2}{w_y^5}\right)n_{pl} - d_{LT}\right] \geq 0.6\sqrt{\frac{w_y}{w_z}}\frac{W_{el,y}}{W_{pl,y}}$$

$$\text{with } d_{LT} = 2a_{LT}\frac{\bar{\lambda}_0}{0.1+\bar{\lambda}_z^4}\frac{M_{y,Ed}}{C_{my}\chi_{LT}M_{pl,y,Rd}}\frac{M_{z,Ed}}{C_{mz}M_{pl,z,Rd}}$$

$$C_{zz} = 1 + (w_z - 1)\left[\left(2 - \frac{1.6}{w_z}C_{mz}^2\bar{\lambda}_{max} - \frac{1.6}{w_z}C_{mz}^2\bar{\lambda}_{max}^2\right)n_{pl} - e_{LT}\right] \geq \frac{W_{el,z}}{W_{pl,z}}$$

$$\text{with } e_{LT} = 1.7a_{LT}\frac{\bar{\lambda}_0}{0.1+\bar{\lambda}_z^4}\frac{M_{y,Ed}}{C_{my}\chi_{LT}M_{pl,y,Rd}}$$

$$\bar{\lambda} = \max\begin{cases}\bar{\lambda}_y \\ \bar{\lambda}_z\end{cases}$$

$\bar{\lambda}_0 = $ *non-dimensional slenderness for lateral-torsional buckling due to uniform bending moment,*
   i.e. $\psi = 1.0$ in Table A.2

$\bar{\lambda}_{LT} = $ *non-dimensional slenderness for lateral-torsional buckling*

If $\bar{\lambda}_0 \leq 0.2\sqrt{C_l}\sqrt[4]{\left(1-\dfrac{N_{Ed}}{N_{cr,z}}\right)\left(1-\dfrac{N_{Ed}}{N_{cr,T}}\right)}$:

$C_{my} = C_{my,0}$

$C_{mz} = C_{mz,0}$

$C_{mLT} = 1.0$

If $\bar{\lambda}_0 > 0.2\sqrt{C_l}\sqrt[4]{\left(1-\dfrac{N_{Ed}}{N_{cr,z}}\right)\left(1-\dfrac{N_{Ed}}{N_{cr,T}}\right)}$:

$C_{my} = C_{my,0} + (1-C_{my,0})\dfrac{\sqrt{\varepsilon_y}a_{LT}}{1+\sqrt{\varepsilon_y}a_{LT}}$

$C_{mz} = C_{mz,0}$

$C_{mLT} = C_{my}^2\dfrac{a_{LT}}{\sqrt{\left(1-\dfrac{N_{Ed}}{N_{cr,z}}\right)\left(1-\dfrac{N_{Ed}}{N_{cr,T}}\right)}} \geq 1$

Table 8.1. Continued

$$\varepsilon_y = \frac{M_{y,Ed}}{N_{Ed}} \frac{A}{W_{el,y}} \quad \text{for class 1. 2 and 3 cross-sections}$$

$$\varepsilon_y = \frac{M_{y,Ed}}{N_{Ed}} \frac{A_{eff}}{W_{eff,y}} \quad \text{for class 4 cross-sections}$$

$N_{cr,y}$ = elastic flexural buckling force about the $y-y$ axis

$N_{cr,z}$ = elastic flexural buckling force about the $z-z$ axis

$N_{cr,T}$ = elastic torsional buckling force

I_T = St Venant torsional constant

I_y = second moment of area about $y-y$ axis

Table 8.2. Equivalent uniform moment factors $C_{mi,0}$ (*Table A.2 of EN 1993-1-1*)

Moment diagram	$C_{mi,0}$
M_1 ▭ ψM_1 $-1 \leq \psi \leq 1$	$C_{mi,0} = 0.79 + 0.21\psi_i + 0.36(\psi_i - 0.33)\dfrac{N_{Ed}}{N_{cr,i}}$
$M(x)$ $M(x)$	$C_{mi,0} = 1 + \left(\dfrac{\pi^2 E I_i |\delta_x|}{L^2 |M_{i,Ed}(x)|} - 1 \right) \dfrac{N_{Ed}}{N_{cr,i}}$ $M_{i,Ed(x)}$ is the maximum moment $M_{y,Ed}$ or $M_{z,Ed}$ $|\delta_x|$ is the maximum member displacement along the member
	$C_{mi,0} = 1 - 0.18\dfrac{N_{Ed}}{N_{cr,i}}$ $C_{mi,0} = 1 + 0.03\dfrac{N_{Ed}}{N_{cr,i}}$

REFERENCE

Boissonnade N, Jaspart J-P, Muzeau J-P and Villette M (2002) Improvement of the interaction formulae for beam columns in Eurocode 3. *Computers and Structures*, **80**, 2375–2385.

Designers' Guide to Eurocode 3: Design of Steel Buildings, 2nd ed.
ISBN 978-0-7277-4172-1

ICE Publishing: All rights reserved
doi: 10.1680/dsb.41721.109

Chapter 9

Annex B (informative) – Method 2: interaction factors k_{ij} for interaction formula in *clause 6.3.3(4)*

As described in the previous chapter, for uniform members subjected to combined bending and axial compression, *clause 6.3.3(4)* provides two interaction formulae, both of which must be satisfied. Each of the interaction formulae contains two interaction factors: k_{yy} and k_{yz} for *equation (6.61)* and k_{zy} and k_{zz} for *equation (6.62)*. Two alternative methods to determine these four interaction factors (k_{yy}, k_{yz}, k_{zy} and k_{zz}) are given by EN 1993-1-1; Method 1 is contained within *Annex A*, and described in the previous chapter, and Method 2 is contained within *Annex B*, and described in this chapter.

Clause 6.3.3(4)

Method 2 is more straightforward than Method 1, and is generally more user-friendly. The background to the method, developed at the Technical Universities of Graz and Berlin, has been described in Lindner (2003).

Clause NA.3.2 of the UK National Annex allows the use of *Annex B*, but states that for section types other than I-, H- and hollow sections, the benefits of plastic redistribution should be neglected – i.e. Class 1 and Class 2 sections should be designed as if they were Class 3 sections. Furthermore, when designing sections that are not doubly symmetric, consideration should be given to the possibility of torsional and torsional-flexural buckling – see Section 13.7 of this guide.

Clause NA.3.2

The basic formulations for determining the interaction factors using Method 2 are given in Table 9.1 (*Table B.1* of EN 1993-1-1) for members not susceptible to lateral–torsional buckling, and in Table 9.2 (*Table B.2* of EN 1993-1-1) for members that are susceptible to lateral–torsional buckling.

The equivalent uniform moment factors C_{my}, C_{mz} and C_{mLT} may be determined from Table 9.3 (*Table B.3* of EN 1993-1-1). C_{my} relates to in-plane major axis bending; C_{mz} relates to in-plane minor axis bending; and C_{mLT} relates to out-of-plane buckling. When referring to Table 9.3 (*Table B.3* of EN 1993-1-1):

- for no loading between points of restraint, the top row of Table 9.3 applies, which gives $C_{mi} = 0.6 + 0.4\psi$ (but a minimum value of 0.4 is prescribed)
- for uniform loading between restraints (indicated by unbroken lines in the moment diagrams), the second and third rows of Table 9.3 apply, with C_{mi} factors derived from the left-hand section of the final column
- for concentrated loading between restraints (indicated by dashed lines in the moment diagrams), the second and third rows of Table 9.3 apply, with C_{mi} factors derived from the right-hand section of the final column.

In structures that rely on the flexural stiffness of the columns for stability (i.e. unbraced frames), Table 9.3 (*Table B.3* of EN 1993-1-1) indicates that the equivalent uniform moment factor (C_{my} or C_{mz}) should be taken as 0.9.

Table 9.1. Interaction factors k_{ij} for members not susceptible to torsional deformations (*Table B.1* of EN 1993-1-1)

Interaction factors	Type of sections	Design assumptions	
		elastic cross-sectional properties class 3, class 4	plastic cross-sectional properties class 1, class 2
k_{yy}	I-sections RHS-sections	$C_{my}\left(1 + 0.6\bar{\lambda}_y \dfrac{N_{Ed}}{\chi_y N_{Rk}/\gamma_{M1}}\right)$ $\leq C_{my}\left(1 + 0.6\dfrac{N_{Ed}}{\chi_y N_{Rk}/\gamma_{M1}}\right)$	$C_{my}\left(1 + (\bar{\lambda}_y - 0.2)\dfrac{N_{Ed}}{\chi_y N_{Rk}/\gamma_{M1}}\right)$ $\leq C_{my}\left(1 + 0.8\dfrac{N_{Ed}}{\chi_y N_{Rk}/\gamma_{M1}}\right)$
k_{yz}	I-sections RHS-sections	k_{zz}	$0.6k_{zz}$
k_{zy}	I-sections RHS-sections	$0.8k_{yy}$	$0.6k_{yy}$
k_{zz}	I-sections	$C_{mz}\left(1 + 0.6\bar{\lambda}_z \dfrac{N_{Ed}}{\chi_z N_{Rk}/\gamma_{M1}}\right)$ $\leq C_{mz}\left(1 + 0.6\dfrac{N_{Ed}}{\chi_z N_{Rk}/\gamma_{M1}}\right)$	$C_{mz}\left(1 + (2\bar{\lambda}_z - 0.6)\dfrac{N_{Ed}}{\chi_z N_{Rk}/\gamma_{M1}}\right)$ $\leq C_{mz}\left(1 + 1.4\dfrac{N_{Ed}}{\chi_z N_{Rk}/\gamma_{M1}}\right)$
	RHS-sections		$C_{mz}\left(1 + (\bar{\lambda}_z - 0.2)\dfrac{N_{Ed}}{\chi_z N_{Rk}/\gamma_{M1}}\right)$ $\leq C_{mz}\left(1 + 0.8\dfrac{N_{Ed}}{\chi_z N_{Rk}/\gamma_{M1}}\right)$

For I- and H-sections and rectangular hollow sections under axial compression and uniaxial bending $M_{y,Ed}$ the coefficient k_{zy} may be $k_{zy} = 0$.

Table 9.2. Interaction factors k_{ij} for members susceptible to torsional deformations (*Table B.2* of EN 1993-1-1)

Interaction factors	Design assumptions	
	elastic cross-sectional properties class 3, class 4	plastic cross-sectional properties class 1, class 2
k_{yy}	k_{yy} from Table B.1	k_{yy} from Table B.1
k_{yz}	k_{yz} from Table B.1	k_{yz} from Table B.1
k_{zy}	$\left[1 - \dfrac{0.05\bar{\lambda}_z}{(C_{mLT} - 0.25)}\dfrac{N_{Ed}}{\chi_z N_{Rk}/\gamma_{M1}}\right]$ $\geq \left[1 - \dfrac{0.05}{(C_{mLT} - 0.25)}\dfrac{N_{Ed}}{\chi_z N_{Rk}/\gamma_{M1}}\right]$	$\left[1 - \dfrac{0.1\bar{\lambda}_z}{(C_{mLT} - 0.25)}\dfrac{N_{Ed}}{\chi_z N_{Rk}/\gamma_{M1}}\right]$ $\geq \left[1 - \dfrac{0.1}{(C_{mLT} - 0.25)}\dfrac{N_{Ed}}{\chi_z N_{Rk}/\gamma_{M1}}\right]$ for $\bar{\lambda}_z < 0.4$: $k_{zy} = 0.6 + \bar{\lambda}_z \leq 1 - \dfrac{0.1\bar{\lambda}_z}{(C_{mLT} - 0.25)}\dfrac{N_{Ed}}{\chi_z N_{Rk}/\gamma_{M1}}$
k_{zz}	k_{zz} from Table B.1	k_{zz} from Table B.1

Table 9.3. Equivalent uniform moment factors C_m in *Tables B.1* and *B.2* (*Table B.3* of EN 1993-1-1)

Moment diagram	range		C_{my} and C_{mz} and C_{mLT}	
			uniform loading	concentrated load
M ⬦ ψM	$-1 \leq \psi \leq 1$		$0.6 + 0.4\psi \geq 0.4$	
M_h M_s ψM_h $\alpha_s = M_s/M_h$	$0 \leq \alpha_s \leq 1$	$-1 \leq \psi \leq 1$	$0.2 + 0.8\alpha_s \geq 0.4$	$0.2 + 0.8\alpha_s \geq 0.4$
	$-1 \leq \alpha_s < 0$	$0 \leq \psi \leq 1$	$0.1 - 0.8\alpha_s \geq 0.4$	$-0.8\alpha_s \geq 0.4$
		$-1 \leq \psi < 0$	$0.1(1 - \psi) - 0.8\alpha_s \geq 0.4$	$0.2(-\psi) - 0.8\alpha_s \geq 0.4$
M_h M_s ψM_h $\alpha_h = M_h/M_s$	$0 \leq \alpha_h \leq 1$	$-1 \leq \psi \leq 1$	$0.95 + 0.05\alpha_h$	$0.90 + 0.10\alpha_h$
	$-1 \leq \alpha_h < 0$	$0 \leq \psi \leq 1$	$0.95 + 0.05\alpha_h$	$0.90 + 0.10\alpha_h$
		$-1 \leq \psi < 0$	$0.95 + 0.05\alpha_h(1 + 2\psi)$	$0.90 - 0.10\alpha_h(1 + 2\psi)$

For members with sway buckling mode the equivalent uniform moment factor should be taken $C_{my} = 0.9$ or $C_{mz} = 0.9$ respectively.

C_{my}, C_{mz} and C_{mLT} shall be obtained according to the bending moment diagram between the relevant braced points as follows:

moment factor	bending axis	points braced in direction
C_{my}	y–y	z–z
C_{mz}	z–z	y–y
C_{mLT}	y–y	y–y

REFERENCE

Lindner J (2003) Design of beams and beam columns. *Progress in Structural Engineering* and Materials, **5**, 38–47.

Designers' Guide to Eurocode 3: Design of Steel Buildings, 2nd ed.
ISBN 978-0-7277-4172-1

ICE Publishing: All rights reserved
doi: 10.1680/dsb.41721.113

Chapter 10
Annex AB (informative) – additional design provisions

Annex AB of EN 1993-1-1 is split into two short sections containing additional information for taking account of material non-linearities in structural analysis and simplified provisions for the design of continuous floor beams. It is noted that this annex is intended to be transferred to EN 1990 in future revisions of the codes.

Sections 10.1 and 10.2 of this guide relate to *clauses AB.1* and *AB.2* of EN 1993-1-1, respectively.

Clause AB.1
Clause AB.2

10.1. Structural analysis taking account of material non-linearities
Clause AB.1 states that, in the case of material non-linearities, the action effects in a structure (i.e. the internal members forces and moments) may be determined using an incremental approach. Additionally, for each relevant design situation (or combination of actions), each permanent and variable action should be increased proportionally.

Clause AB.1

10.2. Simplified provisions for the design of continuous floor beams
Clause AB.2 provides two simplified loading arrangements for the design of continuous floor beams with slabs in buildings. The guidance is applicable when uniformly distributed loads are dominant, but may not be applied where cantilevers are present.

Clause AB.2

The two loading arrangements to be considered are as follows:

1. for maximum sagging moments, alternative spans carrying the design permanent and variable loads, and other spans carrying only the design permanent load
2. for maximum hogging moments, any two adjacent spans carrying the design permanent and variable loads, all other spans carrying only the design permanent load.

Designers' Guide to Eurocode 3: Design of Steel Buildings, 2nd ed.
ISBN 978-0-7277-4172-1

ICE Publishing: All rights reserved
doi: 10.1680/dsb.41721.115

Chapter 11
Annex BB (informative) – buckling of components of buildings structures

This chapter is concerned with the supplementary guidance given in EN 1993-1-1 for the buckling of components of buildings, as covered in *Annex BB*. Sections 11.1, 11.2 and 11.3 of this guide relate to *clauses BB.1*, *BB.2* and *BB.3* of EN 1993-1-1, respectively.

Clause BB.1
Clause BB.2
Clause BB.3

Annex BB provides specific guidance on three aspects of member stability for use when determining the resistance of individual members acting as parts of a frame structure:

- buckling lengths for chord or web members in triangulated and lattice structures – L_{cr} values
- stiffness requirements for trapezoidal sheeting to fully restrain a beam against lateral–torsional instability – S or $C_{v,k}$ values
- maximum stable lengths between adjacent lateral or torsional restraints for members containing plastic hinges – L_m or L_k values.

While the first and third of these provisions will be familiar to those used to BS 5950: Part 1, the material on sheeting restraint is new.

11.1. Flexural buckling of members in triangulated and lattice structures

Clause BB.1 provides L_{cr} values for a series of situations covering structures composed of either angle or hollow section members. For the former it largely follows the BS 5950: Part 1 approach of combining end restraint and the effect of eccentricities in the line of force transfer into a single design provision, i.e. the recommended L_{cr} values also recognize the presence of eccentricities and include allowances for it so that the members can be designed as if axially loaded. Behaviour both in the plane of the truss and out of plane are covered, with some recognition being taken of the rotational restraint available to either brace or chord members when the adjacent components possess greater stiffness. In all cases, more competitive values, i.e. smaller buckling lengths, may be used when these can be justified on the basis of either tests or a more rigorous analysis. Further guidance is given in Chapter 20 of *The Steel Designers' Manual* (Davison and Owens, 2011).

Clause BB.1

11.2. Continuous restraints

Expressions are provided for both the shear stiffness S and the torsional restraint stiffness $C_{v,k}$ in terms of beam properties such that the sheeting may be assumed to provide full lateral or torsional restraint, with the result that χ_{LT} may be taken as 1.0 in *equation (6.55)*. The determination of appropriate values of S and $C_{v,k}$ for a particular arrangement should follow the provisions given in Part 1.3 of Eurocode 3.

Although reference is made to Part 1.3 (through *clause BB.2.1*), that document does not provide explicit guidance on the determination of appropriate values of S for particular arrangements of sheeting and fastening. Thus, reference to other sources is necessary, for example Bryan and Davies (1982).

Clause BB.2.1

In contrast, clause 10.1.5.2 of Part 1.3 sets out a detailed procedure for calculating the total rotational stiffness C_D as a combination of the flexural stiffness of the sheeting and the rotational stiffness of the interconnection between the sheeting and the beam. However, the specific formulae for this latter effect presume the beam to be a light purlin with appropriate sheet/purlin fastening arrangements. It is therefore suggested that these need to be applied with caution when considering arrangements of different proportions, e.g. sheeting supported by hot-rolled beams.

It is, of course, possible for sheeting to provide a combination of both lateral and torsional restraint. This has been studied, and a design procedure developed (Nethercot and Trahair, 1975), but it is not covered explicitly by Eurocode 3.

11.3. Stable lengths of segment containing plastic hinges for out-of-plane buckling

The use of plastic design methods requires that the resistance of the structure be governed by the formation of a plastic collapse mechanism. Premature failure due to any form of instability must therefore be prevented. It is for this reason that only cross-sections whose proportions meet the Class 1 limits may be used for members required to participate in a plastic hinge action. Similarly, member buckling must not impair the ability of such members to deliver adequate plastic hinge rotation. Thus, restrictions on the slenderness of individual members are required. Limits covering a variety of conditions are provided in this section:

- stable lengths of uniform members subject to axial compression and uniform moment between adjacent lateral restraints – L_m
- stable lengths of uniform members subject to axial compression and either constant, linearly varying or non-linearly varying moment between points of torsional restraint – L_k or L_s

Figure 11.1. Member with a three flange haunch. 1, tension flange; 2, elastic section (see *Clause 6.3*); 3, plastic stable length (see *Clause BB.3.2.1*) or elastic (see *Clause 6.3.5.3(2)B*); 4, plastic stable length (see *Clause BB.3.1.1*); 5, elastic section (see *Clause 6.3*); 6, plastic hinge; 7, restraints; 8, bending moment diagram; 9, compression flange; 10, plastic stable length (see *Clause BB.3.2*) or elastic (see *Clause 6.3.5.3(2)B*); 11, plastic stable length (see *Clause BB.3.1.2*); 12, elastic section (see *Clause 6.3*), χ and χ_{LT} from N_{cr} and M_{cr}, including tension flange restraint

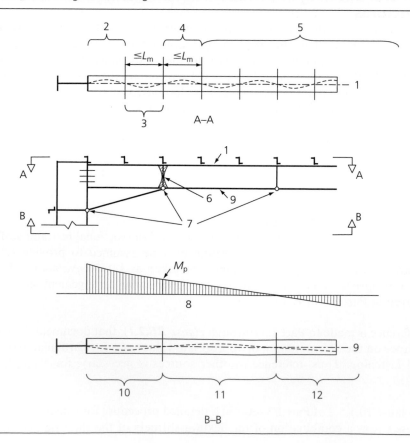

- stable lengths of haunched or tapered members between adjacent lateral restraints – L_m
- stable lengths of haunched or tapered members between torsional restraints – L_s.

In addition, modification factors are given to allow for the presence of a continuous lateral restraint along the tension flange of both uniform and non-uniform members subjected to either linear or non-linear moment gradients.

Figure 11.1 illustrates the nature of the situation under consideration. The overall design premise of failure of the frame being due to the formation of the plastic collapse mechanism requires that a plastic hinge forms as shown at the toe of the rafter (point 6 in Figure 11.1). Both the haunch between this point and the rafter–column joint and the length of the rafter on the opposite side of the plastic hinge extending to the braced location 7 must not fail prematurely by lateral–torsional buckling. For the haunch, depending on the precise conditions of restraint assumed at the brace locations (lateral or torsional) and whether the haunch has three flanges (as illustrated in Figure 11.1) or only two, the maximum stable length may be obtained from one of *equations* (*BB.9*) to (*BB.12*). If the additional benefit of continuous tension flange restraint is to be included then the modification of either *equation* (*BB.13*) or (*BB.14*) should be added. For the uniform length rafter between points 6 and 7, *equations* (*BB.5*) to (*BB.8*), as appropriate, should be used.

Much of this material is very similar to the treatment of the same topic in Appendix G of BS 5950: Part 1. It is, of course, aimed principally at the design of pitched roof portal frame structures, especially when checking stability of the haunched rafter in the eaves region. Further guidance may be found in Gardner (2011) and in Chapters 17 and 19 of *The Steel Designers' Manual* (Davison and Owens, 2011).

REFERENCES

Bryan ER and Davies JM (1982) *Manual of Stressed Skin Diaphragm Design*. London: Granada.

Davison B and Owens GW (2011) *The Steel Designers' Manual*, 7th edn. Steel Construction Institute, Ascot, and Blackwell, Oxford.

Gardner L (2011) *Steel Building Design: Stability of Beams and Columns*. Steel Construction Institute, Ascot, P360.

Nethercot DA and Trahair NS (1975) Design of diaphragm braced I-beams. *Journal of Structural Engineering of the ASCE*, **101**, 2045–2061.

Designers' Guide to Eurocode 3: Design of Steel Buildings, 2nd ed.
ISBN 978-0-7277-4172-1

ICE Publishing: All rights reserved
doi: 10.1680/dsb.41721.119

Chapter 12
Design of joints

12.1. Background

This chapter concerns the subject of joint design, which is covered in EN 1993-1-8 – *Design of Joints*. The purpose of this chapter is to provide an overview of joint design, broadly equivalent to the level of detail on joints provided by BS 5950: Part 1. Unlike the preceding chapters of this guide, where the section numbers in the guide correspond directly to those in Part 1.1 of the code, section numbers in this chapter do not relate to the code.

Part 1.8 of Eurocode 3 is some 50% longer than the general Part 1.1. It provides a much more extensive treatment of the whole subject area of connections than a UK designer would expect to find in a code. At first sight, it is not easy to assimilate the material and to identify which parts are required for which specific application. Traditional UK practice has been for the code document to concern itself only with certain rather fundamental material, e.g. the strengths of bolts and welds, or information on recommended geometrical requirements; EN 1993-1-8 strays into areas more traditionally associated with supplementary design guides. Thus, of its 130 pages, only approximately one-third covers general matters – with the remainder being divided approximately equally between application rules for joints between I-sections and joints between tubular members.

Essentially, the coverage of EN 1993-1-8 focuses on four topics:

- Fasteners (Sections 3 and 4 of EN 1993-1-8), covering the basic strength of bolts in shear, the resistance of fillet welds, etc.
- The role of connections in overall frame design (Section 5 of EN 1993-1-8), covering the various possible approaches to joint classification and global frame analysis.
- Joints between I-sections (Section 6 of EN 1993-1-8), being more akin to the BCSA/SCI 'Green Books' (BCSA/SCI, 1995, 2011) treatment than to the current content of BS 5950: Part 1.
- Joints between structural hollow sections (Section 7 of EN 1993-1-8), being very similar to several existing CIDECT guides (Wardenier *et al.*, 2008; Packer *et al.*, 2009).

It seems likely that established design guides such as the 'Green Book' series (a Eurocode version of which has now been published), design software and other specialist material will emerge to support the technical content of this document. An ECCS inspired document dealing with 'simple connections' is available (Jaspart *et al.*, 2009). This chapter directs the reader to the most generally relevant material and provides some interpretation and guidance on aspects of the more detailed content of EN 1993-1-8.

12.2. Introduction

Section 1 of EN 1993-1-8 covers the scope of the document and provides a useful list of definitions (clause 1.3 of EN 1993-1-8) and symbols (clause 1.4 of EN 1993-1-8). The latter is more than just the usual list of notation due to the need to define numerous geometrical parameters associated with the detailed arrangements for various forms of joint.

12.3. Basis of design

In Section 2 of EN 1993-1-8, partial factors γ_M for the various components present in joints are listed (see Table 2.1 of EN 1993-1-8), of which the most common are:

- resistance of bolts, pins, welds and plates in bearing – γ_{M2}

Table 12.1. Numerical values of partial factors γ_M relevant to connections

Partial factor, γ_M	Eurocode 3 and UK NA
γ_{M2}	1.25
γ_{M3}	1.25 (or 1.1 for serviceability)
γ_{M5}	1.0

- slip resistance – γ_{M3}
- resistance of joints in hollow section lattice girders – γ_{M5}.

The numerical values for these partial factors, as defined by Eurocode 3 and the UK National Annex to EN 1993-1-8, are given in Table 12.1.

Either linear elastic or elastic–plastic analysis may be used to determine the forces in the component parts of a joint subject to the set of basic design concepts listed in clause 2.5 of EN 1993-1-8. These accord with the usual principles adopted when designing joints (Owens and Cheal, 1988). The effects of eccentricity in the line of action of forces should be allowed for using the principles listed in clause 2.7 of EN 1993-1-8.

Section 2 of EN 1993-1-8 concludes with an extensive list of reference standards covering the usual components typically found in joints, e.g. bolts, nuts and washers, or needed to construct joints, e.g. welding consumables.

12.4. Connections made with bolts, rivets or pins

12.4.1 General

Table 3.1 of EN 1993-1-8 lists five grades of bolts, ranging from 4.6 to 10.9 and including the UK norm of 8.8. Only appropriate grade 8.8 or 10.9 bolts may be designed as preloaded.

Three situations for bolts designed to operate in shear are defined in clause 3.4.1 of EN 1993-1-8:

- bearing type – the most usual arrangement
- slip-resistant at serviceability limit state – ultimate condition governed by strength in shear or bearing
- slip-resistant at ultimate limit state – ultimate condition governed by slip.

Similarly, two categories for bolts used in tension are defined:

- non-preloaded – the most usual category
- preloaded – when controlled tightening is employed.

Table 3.2 of EN 1993-1-8 lists the design checks needed for each of these above five arrangements.

Information on geometrical restrictions on the positioning of bolt holes is provided in Table 3.3 of EN 1993-1-8. This generally accords with the provisions of BS 5950: Part 1. It includes the usual provisions for regular and staggered holes in tension members; this topic is covered in Section 6.2.2 of this guide, with reference to the provision of EN 1993-1-1.

12.4.2 Design resistance

Table 3.4 of EN 1993-1-8 lists the design rules for individual bolts subjected to shear and/or tension. For shear, the resistance is given by

$$F_{v,Rd} = \frac{\alpha_v f_{ub} A}{\gamma_{M2}}$$

(D12.1)

where

α_v = 0.6 for classes 4.6, 5.6 and 8.8 where the shear plane passes through the threaded portion of the bolt, and for all classes where the shear plane passes through the unthreaded portion of the bolt

= 0.5 for classes 4.8, 5.8, 6.8 and 10.9 where the shear plane passes through the threaded portion of the bolt

Figure 12.1. Definitions for p_1, e_1, p_2 and e_2

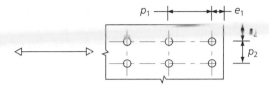

f_{ub} is the ultimate tensile strength of the bolt

A is the tensile stress area when the shear plane passes through the threaded portion of the bolt or the gross cross-sectional area when the shear plane passes through the unthreaded portion of the bolt.

For bearing, the resistance is given by

$$F_{b,Rd} = \frac{k_1 \alpha_b f_u dt}{\gamma_{M2}}$$

(D12.2)

where α_b is the smallest of α_d, f_{ub}/f_u or 1.0, f_u is the ultimate tensile strength of the connected parts, and (with reference to Figure 12.1):

- in the direction of load transfer,

$$\alpha_d = \frac{e_1}{3d_0} \qquad \text{for end bolts}$$

$$\alpha_d = \frac{p_1}{3d_0} - 0.25 \qquad \text{for inner bolts}$$

- perpendicular to the direction of load transfer, k_1 is the smaller of:

$$\left(2.8 \times \frac{e_2}{d_0} - 1.7 \right) \quad \text{or } 2.5 \qquad \text{for edge bolts}$$

$$\left(1.4 \times \frac{p_2}{d_0} - 1.7 \right) \quad \text{or } 2.5 \qquad \text{for inner bolts}$$

The symbols p_1, e_1, p_2 and e_2 are defined in Figure 12.1.

For tension the resistance is

$$F_{t,Rd} = \frac{k_2 f_{ub} A_s}{\gamma_{M2}}$$

(D12.3)

where

A_s is the tensile stress area of the bolt

$k_2 = 0.9$ (except for countersunk bolts, where $k_2 = 0.63$).

For combined shear and tension the resistance is covered by the formula

$$\frac{F_{v,Ed}}{F_{v,Rd}} + \frac{F_{t,Ed}}{1.4F_{t,Rd}} \leq 1.0$$

(D12.4)

Special provisions are made when using oversize or slotted holes or countersunk bolts.

Where bolts transmit load in shear and bearing, and pass through packing of total thickness t_p (Figure 12.2), the design shear resistance should be reduced by a factor β_p, given by

$$\beta_p = \frac{9d}{8d + 3t_p} \qquad \text{but } \beta_p \leq 1.0$$

(D12.5)

Figure 12.2. Fasteners through packing

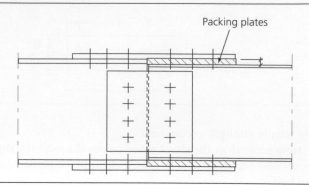

For preloaded bolts the design value of preload $F_{p,Cd}$ is given by

$$F_{p,Cd} = \frac{0.7 f_{ub} A_s}{\gamma_{M7}}$$ (D12.6)

Provisions are also given for injection bolts (clause 3.6.2 of EN 1993-1-8), bolt groups (clause 3.7 of EN 1993-1-8) in bearing and long joints (clause 3.8 of EN 1993-1-8). For long joints, the design shear resistance of all fasteners should be reduced by multiplying by the reduction factor β_{Lf}, given by

$$\beta_{Lf} = 1 - \frac{L_j - 15d}{200d} \qquad \text{but } 0.75 \leq \beta_{Lf} \leq 1.0$$ (D12.7)

where L_j is the distance between the centres of the end bolts in the joint.

12.4.3 Slip-resistant connections

Slip-resistant connections should be designed using the provisions of clause 3.9 of EN 1993-1-8, which gives the design slip resistance as

$$F_{s,Rd} = \frac{k_s n \mu}{\gamma_{M3}} F_{p,C}$$ (D12.8)

where

n is the number of friction surfaces
$F_{p,C} = 0.7 \times 800 A_s$ (subject to conformity with standards).

Values for the factor k_s as well as a set of the slip factor μ corresponding to four classes of plate surface are provided in Tables 3.6 and 3.7 of EN 1993-1-8, respectively.

For situations involving combined tension and shear for which the connection is designed as 'slip-resistant at serviceability', the slip resistance is given by

$$F_{s,Rd,serv} = \frac{k_s n \mu (F_{p,C} - 0.8 F_{t,Ed,serv})}{\gamma_{M3}}$$ (D12.9)

12.4.4 Block tearing

Several cases of block tearing, in which shear failure along one row of bolts in association with tensile rupture along another line of bolts results in the detachment of a piece of material, and thus separation of the connection, are illustrated in Figure 12.3 (Figure 3.8 of EN 1993-1-8).

Equations to cover concentrically and eccentrically loaded situations are provided by equations (D12.10) and (D12.11), respectively:

$$V_{eff,1,Rd} = \frac{f_u A_{nt}}{\gamma_{M2}} + \frac{(1/\sqrt{3}) f_y A_{nv}}{\gamma_{M0}}$$ (D12.10)

Figure 12.3. Block tearing

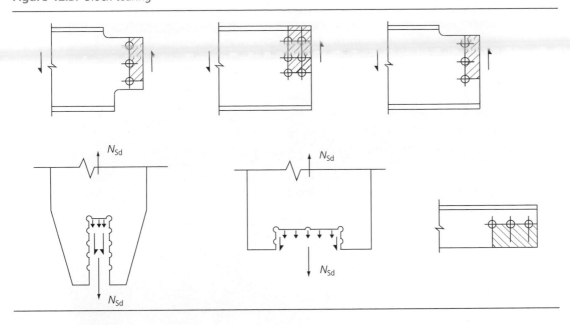

$$V_{\text{eff,2,Rd}} = \frac{0.5 f_u A_{nt}}{\gamma_{M2}} + \frac{(1/\sqrt{3}) f_y A_{nv}}{\gamma_{M0}} \qquad\qquad\text{(D12.11)}$$

where

A_{nt} is the net area subject to tension
A_{nv} is the net area subject to shear.

These equations differ from those of the earlier ENV document in using net area for both shear and tension. Recent work in Canada that paralleled criticism of the American Institute of Steel Construction (AISC) treatment (Driver *et al.*, 2004) has suggested that the original ENV concept was both physically more representative of the behaviour obtained in tests and gave clear yet still safe side predictions of the relevant experimented data.

Both variants are conceptually similar to the treatment given in BS 5950: Part 1, although the form of presentation is different.

Rules are also provided for the tensile resistance of angles connected through one leg that adopt the usual practice of treating it as concentrically loaded but with a correction factor applied to the area.

12.4.5 Prying forces

Although clause 3.11 of EN 1993-1-8 specifically requires that prying forces in bolts loaded in tension be allowed for 'where this can occur', no information on how to recognise such situations or what procedure to use to determine their values is provided. Thus, the interaction equation given in Table 3.4 of EN 1993-1-8 should be treated similarly to the second formula in clause 6.3.4.4 of BS 5950: Part 1. In the absence of specific guidance it seems reasonable to use the procedure of clause 6.3.4.3 of BS 5950: Part 1 to determine the total tensile bolt load $F_{t,Ed}$.

12.4.6 Force distributions at ultimate limit state

A plastic distribution of bolt forces may be used except:

- for connections designed as 'slip-resistant at ultimate'
- when shear (rather than bearing) is the governing condition
- in cases where the connection must be designed to resist the effects of impact, vibration and load reversal (except that caused solely by wind loading).

Figure 12.4. Pin connection

It is stated that any plastic approach is acceptable providing it satisfies equilibrium and that the resistance and ductility of individual bolts is not exceeded.

12.4.7   Connections made with pins

For connections made with pins (Figure 12.4), two cases are recognised:

- where no rotation is required, and the pin may be designed as if it were a single bolt
- all other arrangements for which the procedures given in clause 3.13.2 of EN 1993-1-8 should be followed.

Table 3.10 of EN 1993-1-8 lists the design requirements for pins for shear, bearing (pin and plates), bending and combined shear and bending. A further limit on the contact bearing stress is applied if the pin is to be designed as replaceable. Apart from changes to some of the numerical coefficients, these rules are essentially similar to those in BS 5950: Part 1.

12.5.   Welded connections
12.5.1   General

Design information is provided for welds covering material thicknesses in excess of 4 mm, although for welds in structural hollow sections this limit is reduced to 2.5 mm, with specific guidance being provided in Section 7 of EN 1993-1-8. For thinner materials, reference should normally be made to Part 1.3 of the code. Information on fatigue aspects of weld design is provided in Part 1.9, and on fracture in Part 1.10. It is generally assumed that the properties of the weld metal will be at least the equivalent in terms of strength, ductility and toughness to those of the parent material.

All major types of structural weld are covered, as listed in clause 4.3.1 of EN 1993-1-8.

12.5.2   Fillet welds

The usual geometrical restrictions that the included angle be between 60 and 120° applies, although ways of designing outside these limits are suggested. Intermittent fillet welds must meet the requirements of Figure 4.1 of EN 1993-1-8 in terms of the ratio of hit/miss lengths. A minimum length of 30 mm or six times the throat thickness is required before the weld can be considered as load-carrying. Figure 4.3 of EN 1993-1-8 indicates how the effective weld thickness should be measured; this should not be less than 3 mm. For deep-penetration fillet welds, as defined by Figure 4.4 of EN 1993-1-8, testing is necessary to demonstrate that the required degree of penetration can be achieved consistently.

Two methods are permitted for the design of fillet welds:

■ the directional method, in which the forces transmitted by a unit length of weld are resolved into parallel and perpendicular components
■ the simplified method, in which only longitudinal shear is considered.

These approaches broadly mirror those used in the 2000 and 1990 versions, respectively, of BS 5950: Part 1.

Arguably, the most basic design parameters for fillet welds are effective length and leg length/throat size. For the former, the full length over which the fillet is full size should be used; frequently, this will be the overall length less twice the throat thickness to allow for start/stop end effects.

Directional method
Normal and shear stresses of the form shown in Figure 12.5 (Figure 4.5 of EN 1993-1-8) are assumed, in which:

■ $\sigma_\perp$ is the normal stress perpendicular to the throat
■ $\sigma_\parallel$ is the normal stress parallel to the axis of the throat
■ $\tau_\perp$ is the shear stress perpendicular to the axis of the weld
■ $\tau_\parallel$ is the shear stress parallel to the axis of the weld.

$\sigma_\parallel$ is assumed not to influence the design resistance, while $\sigma_\perp$, $\tau_\perp$ and $\tau_\parallel$ must satisfy the pair of conditions given by equations (D12.12a) and (D12.12b):

$$[\sigma_\perp^2 + 3(\tau_\perp^2 + \tau_\parallel^2)]^{0.5} \le \frac{f_u}{\beta_w \gamma_{M2}} \qquad \text{(D12.12a)}$$

$$\sigma_\perp \le \frac{f_u}{\gamma_{M2}} \qquad \text{(D12.12b)}$$

where

f_u is the nominal ultimate strength of the weaker part joined
β_w is a factor (between 0.8 and 1.0) depending on the steel type (see Table 4.1 of EN 1993-1-8).

Simplified method
At all points along its length, the resultant of all forces per unit length transmitted by the weld ($F_{w,Ed}$) must not exceed the design weld resistance per unit length ($F_{w,Rd}$), where this is simply

Figure 12.5. Stresses on the throat section of a fillet weld

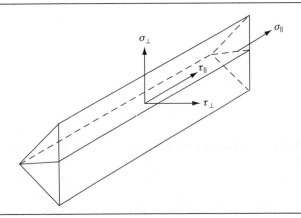

Figure 12.6. Effective width of an unstiffened T joint

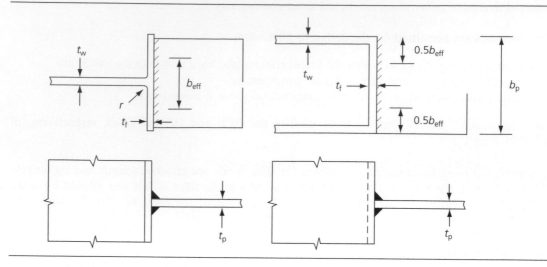

the product of the design shear strength $f_{vw.d}$ and the throat thickness a. The value of $f_{vw.d}$ should be taken as

$$f_{vw.d} = \frac{f_u/\sqrt{3}}{\beta_w \gamma_{M2}} \qquad (D12.13)$$

12.5.3 Butt welds

For full-penetration butt welds the design resistance is simply taken as the strength of the weaker parts connected. This presumes the use of welding consumables that deliver all weld-tensile specimens of greater strength than the parent metal. Partial-penetration butt welds should be designed as deep-penetration fillet welds. Providing the nominal throat thickness of a T-butt weld exceeds the thickness t of the plate forming the stem of the T joint and any unwelded gap does not exceed $t/5$, such arrangements may be designed as if they were full-penetration welds.

12.5.4 Force distribution

Either elastic or plastic methods may be used to determine the distribution of forces in a welded connection. Ductility should be ensured.

12.5.5 Connections to unstiffened flanges

The provisions of this section need to be read in association with the later material in Sections 6 and 7 when dealing with plates attached to I- or H- or rectangular hollow sections. Specific rules are given for the determination of an effective width of plate b_{eff} as defined in Figure 12.6 (Figure 4.8 of EN 1993-1-8) for use in the design expression

$$F_{fc,Rd} = \frac{b_{eff,b,fc} t_{tb} f_{y,fb}}{\gamma_{M0}} \qquad (D12.14)$$

12.5.6 Long joints

Apart from arrangements in which the stress distribution along the weld corresponds to that in the adjacent parts, e.g. web-to-flange girder welds, joints with lengths greater than $150a$ for lap joints or $1.7\,\text{m}$ for joints connecting transverse stiffeners to web plates should be designed by reducing the basic design resistance by a factor β_{Lw} given respectively by

$$\beta_{Lw.1} = 1.2 - 0.2 L_j/150a \qquad \text{but } \beta_{Lw.1} \leq 1.0 \qquad (D12.15)$$

and

$$\beta_{Lw.2} = 1.1 - L_w/17 \qquad \text{but } \beta_{Lw.2} \leq 1.0 \beta_{Lw.2} \geq 0.6 \qquad (D12.16)$$

where

L_j is the overall length of the lap in the direction of the force transfer (in metres)
L_w is the length of the weld (in metres).

12.5.7   Angles connected by one leg

Good practice rules are provided to define situations for which tension at the root of a weld must be explicitly considered and for determining the effective area for angles connected by only one leg so that they may be treated as concentrically loaded. In both cases the provisions are essentially similar to normal UK practice.

12.6.   Analysis, classification and modelling
12.6.1   Global analysis

Readers accustomed to the rather cursory linkage between the properties of joints and their influence on the performance of a structure provided in BS 5950 will be surprised at the level of detail devoted to this topic in Eurocode 3. Although British code BS 5950 and its forerunner BS 449 have always recognised three types of framing,

- simple construction
- semi-rigid construction (termed 'semi-continuous' in Eurocode 3)
- continuous construction,

Eurocode 3 links each type of framing to each of the three methods of global analysis,

- plastic
- rigid–plastic
- elastic–plastic,

in a far more explicit and detailed fashion. It does this via the process of classification of joint types in terms of their strength (moment resistance) and their (rotational) stiffness. Table 12.2 (Table 5.1 of EN 1993-1-8) summarises this process. Central to it is the concept of the moment–rotation characteristic of the joint, i.e. the relationship between the moment the joint can transmit and the corresponding joint rotation. Figure 12.7 illustrates this schematically for a series of idealised joint types.

Clauses 5.1.2 to 5.1.4 of EN 1993-1-8 set out the requirements in terms of joint properties necessary for the use of each of the three types of global analysis. Reading these in association with clause 5.2 of EN 1993-1-8 on the classification of joints permits the following straight-forward options to be identified:

- joints defined as 'nominally pinned', i.e. incapable of transmitting significant moments and capable of accepting the resulting rotations under the design loads – design frame according to the principles of 'simple construction'
- joints defined as 'rigid and full strength', i.e. having sufficient stiffness to justify an analysis based on full continuity and a strength at least equal to that of the connected members – design frame according to the principles of 'continuous construction' using either of elastic, elastic–plastic or rigid–plastic analysis.

Table 12.2. Type of joint model

Method of global analysis		Classification of joint	
Elastic	Nominally pinned	Rigid	Semi-rigid
Rigid–plastic	Nominally pinned	Full strength	Partial strength
Elastic–plastic	Nominally pinned	Rigid and full strength	Semi-rigid and partial strength Semi-rigid and full strength Rigid and partial strength
Type of joint model	Simple	Continuous	Semi-continuous

Figure 12.7. Moment–rotation characteristics of joints

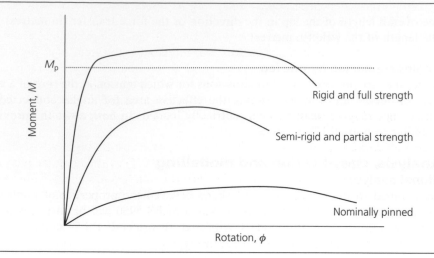

For stiffness, clause 5.2.2.1 of EN 1993-1-8 states that joint classification may be on the basis of one of:

- experimental evidence
- experience of previous satisfactory performance
- calculation.

Interestingly, the equivalent clause for joint strength, clause 5.2.3.1 of EN 1993-1-8, does not contain similar wording, and thus might be interpreted as allowing only the calculation-based approach, i.e. comparing its design moment resistance $M_{j,Rd}$ with the design moment resistance of the members it connects. Given the amount of attention devoted to improving the design of both 'simple' and 'moment' connections in the UK during the past 15 years and the volume of underpinning knowledge (Nethercot, 1998) of the actual behaviour of the types commonly used in the UK embodied within the BCSA/SCI Green Books (BCSA/SCI, 1995, 2011) it is reasonable to presume that 'experience of previous satisfactory performance' would also be accepted as the basis for classifying these types as either nominally pinned or full strength.

Clause 5.1.5 of EN 1993-1-8 provides a similarly detailed treatment of secondary moments caused by the rotational stiffness of the joints and moments resulting from eccentricities and/ or loading between panel points for lattice girders.

Designers wishing to adopt the semi-continuous option should ensure that they are properly acquainted with the subject; this will require study of far more than just the provisions of Eurocode 3. Suitable background texts include those by Anderson (1996) and Faella *et al.* (2000). These texts explain the background to the concept of joint modelling (clause 5.3 of EN 1993-1-8) necessary for the explicit inclusion of joint stiffness and partial strength properties when conducting a frame analysis.

12.7. Structural joints connecting H- or I-sections
12.7.1 General
Section 6 of EN 1993-1-8 explains the principles and application of the concept known as the 'component method'. Since it does this in the context of the determination of

- design moment resistance $M_{j,Rd}$
- rotational stiffener S_j
- rotation capacity φ_{cd}

it is essentially oriented towards semi-continuous construction, i.e. the material has little relevance to joints in simple construction.

The joint is regarded as a rotational spring located at the intersection of the centrelines of the beam and column that it connects and possessing a design moment–rotation relationship. Figure 6.1 of EN 1993-1-8 illustrates the concept: Figure 6.1c of EN 1993-1-8 represents the behaviour that would be expected from a physical test of the arrangement of Figure 6.1a of EN 1993-1-8. In order to obtain the three key measures of performance, $M_{j,Rd}$, S_j and φ_{cd}, the joint is 'broken down' into its basic components, e.g. shear in the web panel of the column, or tension in the bolts, and expressions or calculation procedures for determining its contribution to each of the three performance measures are given in Table 6.1 of EN 1993-1-8. The remaining clauses of Section 6 then define and explain those expressions and procedures.

Readers intending to implement the material of this chapter are strongly advised to prepare themselves by studying the relevant part of the BCSA/SCI (1995) guide on moment connections, since this provides a simplified and more familiar introduction to the subject. Those wishing to design 'nominally pinned' connections for use in structures designed according to the principles of simple construction are therefore advised to use the Eurocode version of the Green Book on simple connections (BCSA/SCI, 2011). The majority of its content will be familiar to those who have used the BS 5950 versions, since changes are essentially only to the formulae and values for fastener strength of the type described at the beginning of this chapter.

12.8. Structural joints connecting hollow sections
12.8.1 General
Section 7 of EN 1993-1-8 covers the design of structural joints connecting hollow sections (Figure 12.8). Readers already familiar with the CIDECT series of design guides for structural hollow sections will find much of Section 7 of EN 1993-1-8 familiar. It is, however, limited to the design of welded connections for static loading, though guidance for fatigue loading does exist elsewhere (CIDECT, 1982). It covers both uniplanar and multiplanar joints, i.e. two and three dimensions, in lattice structures and deals with circular and rectangular hollow section arrangements. It also contains some provisions for uniplanar joints involving combinations of open and closed sections.

In addition to certain geometrical restrictions, the detailed application rules are limited to joints in which the compression elements of all the members are Class 2 or better. Figure 7.1 of EN 1993-1-8 contains all the geometrical arrangements covered, while Figures 7.2 to 7.4 of EN 1993-1-8 illustrate all the potential failure modes. Six specific modes, defined by clause 7.2.2 of EN 1993-1-8, are covered for the cases of both axial load and moment loading in the brace member.

Figure 12.8. Structural joints connecting hollow sections (The London Eye)

Clauses 7.4 to 7.7 of EN 1993-1-8 provide, largely in tabular form, the detailed expressions and procedures for checking the adequacy of each arrangement. Readers intending to implement these would be well advised to first consult the relevant CIDECT material to obtain the basis and background to the specific provisions.

REFERENCES

Anderson D (ed.) (1996) *Semi-rigid Behaviour of Civil Engineering Structural Connections*. COST-C1, Brussels.

BCSA/SCI (1995) *Joints in Steel Construction – Moment Connections*. Steel Construction Institute, Ascot, P207.

BCSA/SCI (2011) *Joints in Steel Construction – Simple Connections*. Steel Construction Institute, Ascot.

CIDECT (1982) *Cidect Monograph No. 7. Fatigue Behaviour of Welded Hollow Section Joints*. Comité International pour le Développement et l'Étude de la Construction Tubulaire, Cologne.

Driver RG, Grondin GY and Kulak GL (2004) A unified approach to design for block shear. In: *Connections in Steel Structures V: Innovative Steel Connections. ECCS-AISC Workshop*, Amsterdam.

Faella C, Piluso V and Rizzano G (2000) *Structural Steel Semi-rigid Connections*. CRC Press, Boca Raton, FL.

Jaspart JP, Demonceau JF, Renkin S and Guillaume ML (2009) *European Recommendations for the Design of Simple Joints in Steel Structures*. European Convention for Constructional Steelwork, Brussels. Publication No. 126.

Nethercot DA (1998) Towards a standardisation of the design and detailing of connections. *Journal of Constructional Steel Research*, **46**: 3–4.

Owens GW and Cheal BD (1988) *Structural Steelwork Connections*. Butterworth, London.

Packer JA, Wardenier J, Zhao X-L, van der Vege GJ and Kurobane Y (2009) *Design Guide for Rectangular Hollow Section (RHS) Joints Under Predominantly Static Loading*, 2nd edn. Comité International pour le Développement et l'Étude de la Construction Tubulaire (CIDECT), Cologne.

Wardenier J, Kurobone Y, Packer JA, van der Vegte GJ and Zhao X-L (2008) *Design Guide for Circular Hollow Section (CHS) Joints Under Predominantly Static Loading*, 2nd edn. Comité International pour le Développement et l'Étude de la Construction Tubulaire (CIDECT), Cologne.

Designers' Guide to Eurocode 3: Design of Steel Buildings, 2nd ed.
ISBN 978-0-7277-4172-1

ICE Publishing: All rights reserved
doi: 10.1680/dsb.41721.131

Chapter 13
Cold-formed design

This chapter concerns the subject of cold-formed member design, which is covered in EN 1993-1-3 – *General Rules: Supplementary Rules for Cold-formed Thin Gauge Members and Sheeting*. The purpose of this chapter is to provide an overview of the behavioural features of cold-formed structural components and to describe the important aspects of the code. Unlike Chapters 1–11 of this guide, where the section numbers in the guide correspond directly to those in Part 1.1 of the code, section numbers in this chapter do not relate to the code.

13.1. Introduction
Cold-formed, thin-walled construction used to be limited to applications where weight savings were of primary concern, such as the aircraft and automotive industries. However, following improvements in manufacturing techniques, corrosion protection, product availability, understanding of the structural response and sophistication of design codes for cold-formed sections, light-gauge construction has become increasingly widespread. Light-gauge sections used in conjunction with hot-rolled steelwork is now commonplace (Figure 13.1).

The use of thin, cold-formed material brings about a number of special design problems that are not generally encountered when using ordinary hot-rolled sections. These include:

■ non-uniform distribution of material properties due to cold-working
■ rounded corners and the calculation of geometric properties
■ local buckling
■ distortional buckling
■ torsional and flexural torsional buckling
■ shear lag

Figure 13.1. Light-gauge (cold-formed) sections in conjunction with hot-rolled steelwork. (Courtesy of Metsec)

- flange curling
- web crushing, crippling and buckling.

These effects, and their codified treatment, will be outlined in the remainder of this chapter. Further general guidance, covering areas such as connections of cold-formed sections, service-ability considerations, modular construction, durability and fire resistance, may be found in Grubb *et al.* (2001), Gorgolewski *et al.* (2001) and Rhodes and Lawson (1992).

13.2. Scope of Eurocode 3, Part 1.3

EN 1993-1-3 is limited in scope by the maximum width-to-thickness ratios set out in Table 13.1. Use of cross-sections with elements exceeding these proportions must be justified by testing. Interestingly, EN 1993-1-3 also states that its provisions are not to be applied to the design of cold-formed circular and rectangular hollow sections; consequently, buckling curves are not provided for such cross-section types, and reference should be made to EN 1993-1-1.

13.3. Material properties

All cold-forming operations that involve plastic deformation result in changes to the basic material properties; essentially producing increased yield strengths, but with corresponding

Table 13.1. Maximum width-to-thickness ratios covered by EN 1993-1-3

Element of cross section	Maximum value
	$b/t \leq 50$
	$b/t \leq 60$ $c/t \leq 50$
	$b/t \leq 90$ $c/t \leq 60$ $d/t \leq 50$
	$b/t \leq 500$
	$45° \leq \phi \leq 90°$ $c/t \leq 500 \sin \phi$

reductions in ductility (ultimate strain). EN 1993-1-3 allows for strength enhancements due to cold-forming by defining an average (enhanced) yield strength f_{ya} that may be used in subsequent calculations in place of the basic yield strength f_{yb} (with some limitations which are discussed later). The EN 1993-1-3 expression for average yield strength is given by

$$f_{ya} = f_{yb} + \frac{knt^2}{A_g}(f_u + f_{yb}) \qquad \text{but } f_{ya} \leq \frac{f_u + f_{yb}}{2} \qquad\qquad (D13.1)$$

where

t is the material thickness (mm)

A_g is the gross cross-sectional area (in square millimetres)

k is a numerical coefficient that depends on the forming process ($k = 7$ for cold-rolling and $k = 5$ for other forming methods)

n is the number of 90° bends in the cross-section with an internal radius less than or equal to five times the material thickness (fractions of 90° bends should be counted as fractions of n).

The code also allows the average yield strength to be determined on the basis of full-scale laboratory testing.

The code states that the average (enhanced) yield strength may not be used for Class 4 cross-sections (where the section is not fully effective) or where the members have been subjected to heat treatment after forming.

13.4. Rounded corners and the calculation of geometric properties

Cold-formed cross-sections contain rounded corners that make calculation of geometric properties less straightforward than for the case of sharp corners. In such cross-sections, EN 1993-1-3 states that notional flat widths b_p (used as a basis for the calculation of effective section properties) should be measured to the midpoints of adjacent corner elements, as shown in Figure 13.2.

For small internal radii the effect of the rounded corners is small and may be neglected. EN 1993-1-3 allows cross-section properties to be calculated based on an idealised cross-section that comprises flat elements concentrated along the mid-lines of the actual elements, as illustrated in Figure 13.3, provided $r \leq 5t$ and $r \leq 0.10b_p$ (where r is the internal corner radius, t is the material thickness and b_p is the flat width of a plane element).

It should be noted that section tables and design software will generally conduct calculations that incorporate rounded corners, so there may be small discrepancies with hand calculations based on the idealised properties. Examples 13.1 and 13.2 show calculation of the gross and effective sections properties of a lipped channel section, based on the idealisations described.

13.5. Local buckling

Local buckling of compression elements is accounted for in EN 1993-1-3 primarily by making reference to Part 1.5 of the code. As explained in Section 6.2.2 of this guide, an effective width approach is adopted, whereby 'ineffective' portions of a cross-section are removed and section properties may be determined based on the remaining effective portions.

For the cases where calculations may utilise an idealised cross-section (see Section 13.4 of this guide), consisting of plane elements with sharp corners (i.e. provided $r \leq 5t$ and $r \leq 0.10b_p$), then the notional flat widths b_p of the plane elements (used in the calculation of effective section properties) may be simply taken as the widths of the idealised elements.

Calculation of the effective section properties due to local buckling of a lipped channel section is demonstrated in Example 13.1. However, the effect of distortional buckling on such a cross-section should also be determined, and this is the subject of Example 13.2.

Figure 13.2. Notional widths of plane elements b_p allowing for corner radii. (a) Mid-point of corner or bend. (b) Notional flat width b_p of plane elements b, c and d. (c) Notional flat width b_p for a web (b_p = slant height s_w). (d) Notional flat width b_p of plane elements adjacent to stiffeners

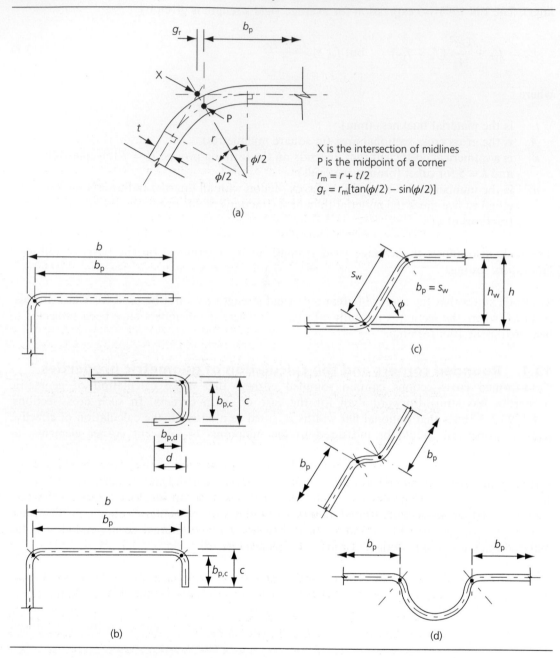

X is the intersection of midlines
P is the midpoint of a corner
$r_m = r + t/2$
$g_r = r_m[\tan(\phi/2) - \sin(\phi/2)]$

Figure 13.3. Idealised cross-section properties

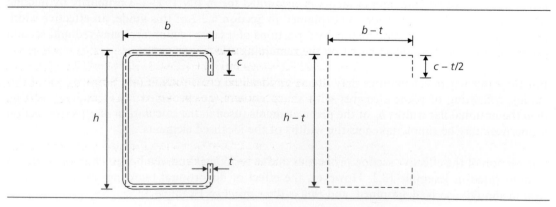

Example 13.1: calculation of section properties for local buckling

Calculate the effective area and the horizontal shift in neutral axis due to local buckling for a $200 \times 65 \times 1.6$ mm lipped channel in zinc-coated steel with a nominal yield strength of 280 N/mm^2 and a Young's modulus of $210\,000$ N/mm^2, and subjected to pure compression. Assume that the zinc coating forms 0.04 mm of the thickness of the section, and ignore the contribution of the coating in the calculations.

Section properties

The section properties are shown in Figure 13.4.

Figure 13.4. Section properties for a $200 \times 65 \times 1.6$ mm lipped channel

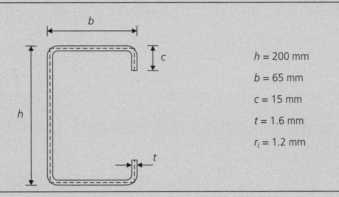

$h = 200$ mm

$b = 65$ mm

$c = 15$ mm

$t = 1.6$ mm

$r_i = 1.2$ mm

The internal corner radii are less than both $5t$ and $0.10b_\mathrm{p}$. Idealised geometry may therefore be adopted without incurring significant errors (EN 1993-1-3).

Idealised cross-section dimensions

The idealised cross-section dimensions are shown in Figure 13.5.

Calculation of gross properties

Gross area

$$A_\mathrm{g} = (198.4 \times 1.56) + (2 \times 63.4 \times 1.56) + (2 \times 14.2 \times 1.56)$$

$$= 551.6 \text{ mm}^2$$

Figure 13.5. Idealised section for the $200 \times 65 \times 1.6$ mm lipped channel

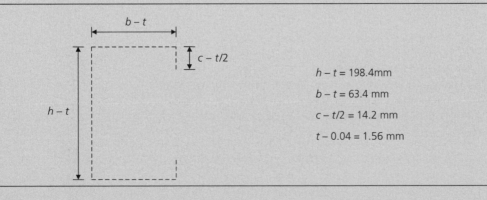

$h - t = 198.4$mm

$b - t = 63.4$ mm

$c - t/2 = 14.2$ mm

$t - 0.04 = 1.56$ mm

The horizontal position of the neutral axis from the centreline of the web for the gross section $\bar{y}_\mathrm{g}$ is

$$\bar{y}_\mathrm{g} = \{[2 \times (63.4 \times 1.56) \times 63.4/2] + [2 \times (14.2 \times 1.56) \times 63.4]\}/551.6$$

$$= 16.46 \text{ mm}$$

Calculation of effective widths
Web:

$k_\sigma = 4.0$ for an internal element in pure compression (EN 1993-1-5, Table 4.1)

$$\varepsilon = \sqrt{235/f_y} = \sqrt{235/280} = 0.92$$

$$\bar{\lambda}_p = \sqrt{\frac{f_y}{\sigma_{cr}}} = \frac{\bar{b}/t}{28.4\varepsilon\sqrt{k_\sigma}} = \frac{198.4/1.56}{28.4 \times 0.92 \times \sqrt{4.0}} = 2.44$$

$$\rho = \frac{\bar{\lambda}_p - 0.055(3 + \psi)}{\bar{\lambda}_p^2} = \frac{2.44 - 0.055 \times (3 + 1)}{2.44^2} = 0.37$$

$$b_{eff} = \rho\bar{b} = 0.37 \times 198.4 = 73.87 \, \text{mm} \qquad \text{(web)}$$

Flanges:

$k_\sigma = 4.0$ for an internal element in pure compression (EN 1993-1-5, Table 4.1)

$$\varepsilon = \sqrt{235/f_y} = \sqrt{235/280} = 0.92$$

$$\bar{\lambda}_p = \sqrt{\frac{f_y}{\sigma_{cr}}} = \frac{\bar{b}/t}{28.4\varepsilon\sqrt{k_\sigma}} = \frac{63.4/1.56}{28.4 \times 0.92 \times \sqrt{4.0}} = 0.78$$

$$\rho = \frac{\bar{\lambda}_p - 0.055(3 + \psi)}{\bar{\lambda}_p^2} = \frac{0.78 - 0.055 \times (3 + 1)}{0.78^2} = 0.92$$
$$b_{eff} = \rho\bar{b} = 0.92 \times 63.4 = 58.31 \, \text{mm} \qquad \text{(flanges)}$$

Lips:

$k_\sigma = 0.43$ for an outstand element in pure compression (EN 1993-1-5, Table 4.2)

$$\varepsilon = \sqrt{235/f_y} = \sqrt{235/280} = 0.92$$

$$\bar{\lambda}_p = \sqrt{\frac{f_y}{\sigma_{cr}}} = \frac{\bar{b}/t}{28.4\varepsilon\sqrt{k_\sigma}} = \frac{14.2/1.56}{28.4 \times 0.92 \times \sqrt{0.43}} = 0.53$$

$$\rho = \frac{\bar{\lambda}_p - 0.188}{\bar{\lambda}_p^2} = \frac{0.53 - 0.188}{0.53^2} = 1.21 \qquad \text{(but } \rho \leq 1\text{)}$$

$$b_{eff} = \rho\bar{b} = 1.00 \times 14.2 = 14.2 \, \text{mm} \qquad \text{(lips)}$$

Gross and effective sections
The gross and effective sections are shown in Figure 13.6.

Calculation of effective section properties
Effective area

$$A_{eff} = (73.87 \times 1.56) + (2 \times 58.31 \times 1.56) + (2 \times 14.2 \times 1.56) = 341.5 \, \text{mm}^2$$

The horizontal position of the neutral axis from the centreline of the web for the effective section is

$$\bar{y}_{eff} = \{[2 \times (29.16 \times 1.56) \times 29.16/2)] + [2 \times (29.16 \times 1.56) \times (63.4 - 29.16/2)]$$

$$+ [2 \times (14.2 \times 1.56) \times 63.4]\}/341.5$$

$$= 25.12 \, \text{mm}$$

Figure 13.6. (a) Gross and (b) effective sections (dimensions in mm)

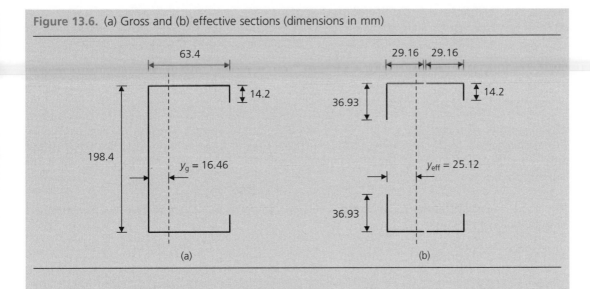

(a)   (b)

The horizontal shift in the neutral axis from the gross section to the effective section is

$$e_{Ny} = 25.12 - 16.46 = 8.66 \text{ mm}$$

Based on the same idealised cross-section, calculations according to BS 5950: Part 5 result in an effective area of 367.2 mm² and a horizontal shift in the neutral axis of 7.75 mm.

13.6. Distortional buckling

13.6.1 Background

Distortional buckling occurs where edge or intermediate stiffeners fail to prevent local displacements of the nodal points (i.e. either at the flange-to-lip junction or at the location of the intermediate stiffeners themselves). Local and distortional buckling modes for cross-sections with edge stiffeners and with intermediate stiffeners are shown in Figures 13.7 and 13.8, respectively.

13.6.2 Outline of the design approach

The EN 1993-1-3 approach to the design of compression elements with edge or intermediate stiffeners, which accounts for distortional buckling, is described herein. The method is based on the assumption that the stiffener behaves as a compression member with continuous partial restraint, represented by a linear spring (of stiffness K). The spring acts at the centroid of the effective stiffener section, as illustrated by Figure 13.9.

13.6.3 Linear spring stiffness K

The stiffness of the linear springs may be derived by means of a unit load analysis, which should include the flexural stiffness of the compression element under consideration and the rotational restraint offered by adjoining elements. Rotational springs, located at the ends

Figure 13.7. (a) Local and (b) distortional buckling of cross-sections containing edge stiffeners

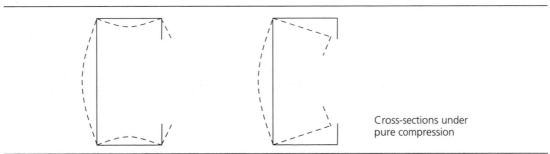

Cross-sections under
pure compression

Figure 13.8. (a) Local and (b) distortional buckling of elements containing intermediate stiffeners

(a) (b)

of the compression element, are employed to reflect the rotational restraint offered by adjoining elements, where the rotational spring stiffness C_θ is dependent upon the flexural stiffness and boundary conditions of the adjoining elements and the stress distribution to which the cross-section is subjected.

The linear spring stiffnesses K for lipped C sections, lipped Z sections and intermediate stiffeners may be taken as follows:

■ For lipped C and lipped Z sections:

$$K_1 = \frac{Et^3}{4(1-\nu^2)} \frac{1}{b_1^2 h_w + b_1^3 + 0.5 b_1 b_2 h_w k_f}$$ (D13.2)

where

b_1 is the distance from the web-to-flange junction to the centroid of the effective edge stiffener section of flange 1
b_2 is the distance from the web-to-flange junction to the centroid of the effective edge stiffener section of flange 2
h_w is the web depth
$k_f = A_{eff2}/A_{eff1}$ for non-symmetric compression
$k_f = 0$ if flange 2 is in tension
$k_f = 1$ for a symmetric section under pure compression
A_{eff1} is the effective area of the edge stiffener section for flange 1
A_{eff2} is the effective area of the edge stiffener section for flange 2.

Flange 1 is the flange under consideration, for which the linear spring stiffness K_1 is being determined, and flange 2 is the opposite flange.

■ For elements with intermediate stiffeners (and conservatively assuming no rotational restraint from the adjoining elements):

$$K = \frac{0.25(b_1 + b_2)Et^3}{(1-\nu^2)b_1^2 b_2^2}$$ (D13.3)

Figure 13.9. Assumed model for edge and intermediate stiffeners. (a) Single-fold edge stiffener. (b) Double-fold edge stiffener. (c) Intermediate stiffener

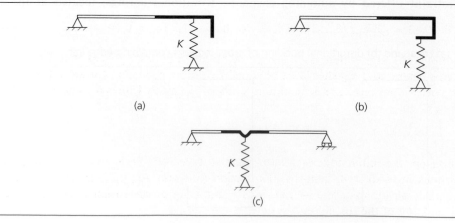

(a) (b)

(c)

Figure 13.10. Initial values of effective widths. (a) Single-edge fold. (b) Double-edge fold

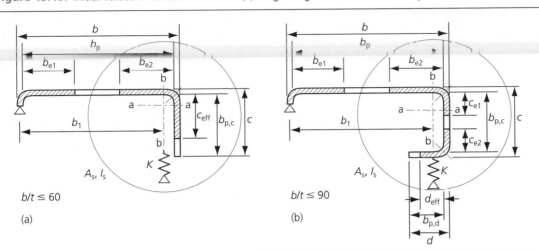

where

b_1 is the distance from the centroid of the intermediate stiffener to one support (web-to-flange junction)

b_2 is the distance from the centroid of the intermediate stiffener to the other support (web-to-flange junction).

13.6.4 Design procedure

The design procedure to obtain an effective section to account for distortional buckling contains three steps.

Step 1

The first step is to calculate the initial effective section (Figure 13.10) for local buckling (i.e. assuming infinite spring stiffness such that the corners or intermediate stiffeners act as nodal points). The maximum compressive stress in the element $\sigma_{\mathrm{com,Ed}}$ should be taken as equal to $f_{\mathrm{yb}}/\gamma_{\mathrm{M0}}$.

Determination of the initial effective widths (b_{e1} and b_{e2}) for doubly supported plane elements should generally be carried out in accordance with EN 1993-1-5, as described in Section 6.2.2 of this guide. Initial values of effective widths for single and double-edge fold stiffeners are treated similarly, except that values of the buckling factor k_σ should be taken as follows:

(a) For determining c_{eff} for a single-fold edge stiffener:

$$k_\sigma = 0.5 \qquad \text{for } b_{\mathrm{p,c}}/b_{\mathrm{p}} \leq 0.35 \tag{D13.4}$$

$$k_\sigma = 0.5 + 0.83 \times \sqrt[3]{[(b_{\mathrm{p,c}}/b_{\mathrm{p}}) - 0.35]^2} \qquad \text{for } 0.35 \leq b_{\mathrm{p,c}}/b_{\mathrm{p}} \leq 0.60 \tag{D13.5}$$

(b) For a double-fold edge stiffener, c_{eff} should be obtained by taking k_σ as that for a doubly supported element, and d_{eff} should be obtained by taking k_σ as that for an outstand element; both values of k_σ are defined in EN 1993-1-5 and in Tables 6.2 and 6.3 of this guide.

Step 2

In the second step, the initial edge or intermediate stiffener section is considered in isolation. The flexural buckling reduction factor of this section (allowing for the linear spring restraint) is then calculated, on the assumption that flexural buckling of the stiffener section represents distortional buckling of the full stiffened element.

The elastic critical buckling stress for the stiffener section is calculated using

$$\sigma_{cr,s} = \frac{2\sqrt{KEI_s}}{A_s}$$

(D13.6)

where

K is the linear spring stiffness, discussed in Section 13.6.3 of this guide
I_s is the second moment of area of the effective stiffener section about its centroidal axis a–a
A_s is the cross-sectional area of the effective stiffener section.

The reduction factor χ_d may hence be obtained using the non-dimensional slenderness $\bar{\lambda}_d$ through the expressions given in equations (D13.7)–(D13.9):

$$\chi_d = 1.0 \quad \text{for } \bar{\lambda}_d \leq 0.65$$

(D13.7)

$$\chi_d = 1.47 - 0.723 \quad \text{for } 0.65 < \bar{\lambda}_d < 1.38$$

(D13.8)

$$\chi_d = 0.66/\bar{\lambda}_d \quad \text{for } \bar{\lambda}_d \geq 1.38$$

(D13.9)

where

$$\bar{\lambda}_d = \sqrt{f_{yb}/\sigma_{cr,s}}$$

Based on the reduction factor χ_d, a reduced area for the effective stiffener section is calculated. The reduced area is calculated from

$$A_{s,red} = \chi_d A_s \frac{f_{yb}/\gamma_{M0}}{\sigma_{com,Ed}} \quad \text{but } A_{s,red} \leq A_s$$

(D13.10)

The reduced area is implemented by means of a uniform reduction in thickness of the effective stiffener sections, as given by

$$t_{red} = t A_{s,red}/A_s$$

(D13.11)

Step 3
The third step is defined in EN 1993-1-3 as optional, but allows the value of χ_d to be refined iteratively using modified values of ρ obtained by taking $\sigma_{com,Ed}$ equal to $\chi_d f_{yb}/\gamma_{M0}$ for each iteration.

The steps are shown for an edge stiffener in Figure 13.11 (from EN 1993-1-3), and are illustrated in Example 13.2.

13.7. Torsional and torsional–flexural buckling
Flexural buckling is the predominant buckling mode for compression members in typical building structures using conventional hot-rolled sections. In light-gauge construction, flexural buckling also governs many design cases, but torsional and torsional–flexural modes must also be checked. The code provisions for flexural buckling in Part 1.3 of the code are essentially the same as those of Part 1.1, although different cross-section types are covered, as shown in Table 13.2.

Torsional buckling is pure twisting of a cross-section, and only occurs in centrally loaded struts which are point symmetric and have low torsional stiffness (e.g. a cruciform section). Torsional–flexural buckling is a more general response that occurs for centrally loaded struts with cross-sections that are singly symmetric and where the centroid and the shear centre do not coincide (e.g. a channel section).

Figure 13.11. Design steps to determine distortional buckling resistance (from EN 1993-1-3)

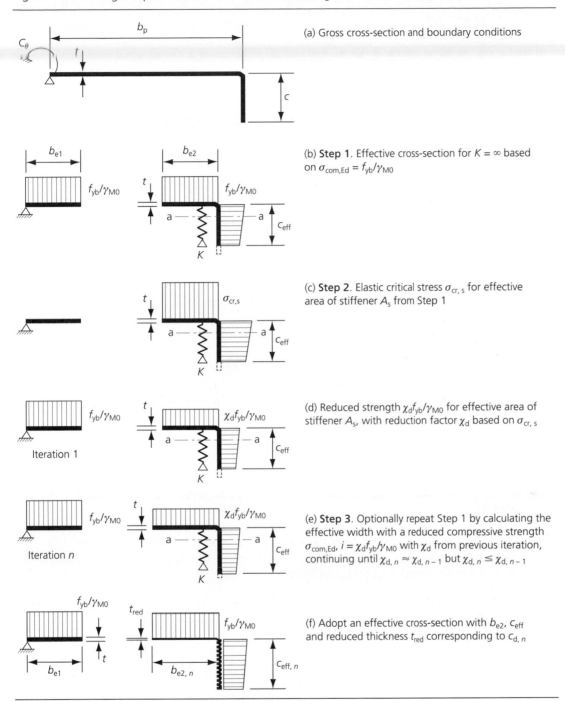

(a) Gross cross-section and boundary conditions

(b) **Step 1**. Effective cross-section for $K = \infty$ based on $\sigma_{com,Ed} = f_{yb}/\gamma_{M0}$

(c) **Step 2**. Elastic critical stress $\sigma_{cr,s}$ for effective area of stiffener A_s from Step 1

(d) Reduced strength $\chi_d f_{yb}/\gamma_{M0}$ for effective area of stiffener A_s, with reduction factor χ_d based on $\sigma_{cr,s}$

(e) **Step 3**. Optionally repeat Step 1 by calculating the effective width with a reduced compressive strength $\sigma_{com,Ed,i} = \chi_d f_{yb}/\gamma_{M0}$ with χ_d from previous iteration, continuing until $\chi_{d,n} \approx \chi_{d,n-1}$ but $\chi_{d,n} \leq \chi_{d,n-1}$

(f) Adopt an effective cross-section with b_{e2}, c_{eff} and reduced thickness t_{red} corresponding to $c_{d,n}$

Example 13.2: cross-section resistance to distortional buckling

This example demonstrates the method set out in EN 1993-1-3 for the calculation of cross-section resistance to (local and) distortional buckling. The example is based on the same $200 \times 65 \times 1.6$ mm lipped channel section of Example 13.1, where effective section properties for local buckling were determined. The idealised cross-section and the properties of the gross cross-section have therefore already been calculated in Example 13.1.

Properties of gross cross-section

$$A_g = 551.6 \text{ mm}^2 \qquad \bar{y}_g = 16.46 \text{ mm}$$

As has been described, the Eurocode method comprises three steps. These steps will be followed in this example.

Step 1: calculation of the initial effective section
For the web and flanges, the initial effective section is as calculated for local buckling in Example 13.1.

Web:

$$b_{eff} = 73.9 \text{ mm}$$

Flanges:

$$b_{eff} = 58.3 \text{ mm}$$

For the lips (single-fold edge stiffeners), the value of the buckling coefficient k_σ should be obtained from equation (D13.4) or (D13.5):

$$b_{p,c}/b_p = 14.2/63.4 = 0.22 \qquad (\leq 0.35)$$

$$\therefore k_\sigma = 0.5$$

$$\varepsilon = \sqrt{235/f_y} = \sqrt{235/280} = 0.92$$

$$\bar{\lambda}_p = \sqrt{\frac{f_y}{\sigma_{cr}}} = \frac{\bar{b}/t}{28.4\varepsilon\sqrt{k_\sigma}} = \frac{14.2/1.56}{28.4 \times 0.92 \times \sqrt{0.50}} = 0.49$$

$$\rho = \frac{\bar{\lambda}_p - 0.188}{\bar{\lambda}_p^2} = \frac{0.49 - 0.188}{0.49^2} = 1.25 \qquad (\text{but } \rho \leq 1)$$

Lips:

$$b_{eff} = \rho\bar{b} = 1.00 \times 14.2 = 14.2 \text{ mm}$$

The initial effective section is therefore as that given for local buckling in Example 13.1 (see Figure 13.6).

Step 2: calculation of reduced thickness for effective edge stiffener section
The effective edge stiffener section shown in Figure 13.12 is now considered in isolation to determine the distortional buckling resistance.

Figure 13.12. Effective edge stiffener section

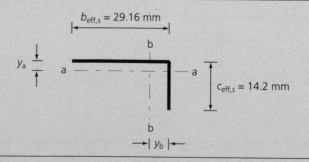

Calculation of geometric properties of effective edge stiffener section
(Symbols defined in Figure 13.12 and in Sections 13.6.3 and 13.6.4 of this guide.)

$$y_a = (14.2 \times 1.56 \times 14.2/2)/[(29.16 + 14.2) \times 1.56] = 2.3 \text{ mm}$$

$$y_b = (29.16 \times 1.56 \times 29.16/2)/[(29.16 + 14.2) \times 1.56] = 9.8 \text{ mm}$$

$$A_s = (29.16 + 14.2) \times 1.56 = 67.6 \text{ mm}^2$$

$$I_s = (29.16 \times 1.56^3)/12 + (1.56 \times 14.2^3)/12 + (29.16 \times 1.56 \times 2.3^2)$$
$$+ [14.2 \times 1.56 \times (14.2/2 - 2.3)^2]$$
$$= 1132.4\,\text{mm}^4$$

Calculation of linear spring stiffness K

From equation (D13.2)

$$K_1 = \frac{Et^3}{4(1-\nu^2)} \frac{1}{b_1^2 h_w + b_1^3 + 0.5 b_1 b_2 h_w k_f}$$

$b_1 = b_2 = 63.4 - 9.8 = 53.6\,\text{mm}$

$k_f = 1.0$ for a symmetric section under pure compression

$\nu = 0.3$

$h_w = 198.4\,\text{mm}$

$\therefore K_1 = 0.22\,\text{N/mm (per unit length)}$

Elastic critical buckling stress for the effective stiffener section

From equation (D13.6)

$$\sigma_{cr,s} = \frac{2\sqrt{KEI_s}}{A_s} = \frac{2\sqrt{0.22 \times 210\,000 \times 1132.5}}{67.6} = 212\,\text{N/mm}^2$$

Reduction factor χ_d for distortional buckling

Non-dimensional slenderness

$$\bar{\lambda}_d = \sqrt{f_{yb}/\sigma_{cr,s}} = \sqrt{280/212} = 1.15$$

$\therefore 0.65 < \bar{\lambda}_d < 1.38$

so, from equation (D13.8),

$\chi_d = 1.47 - 0.723\bar{\lambda}_d = 0.64$

Reduced area (and thickness) of effective stiffener section

$$A_{s,red} = \chi_d A_s \frac{f_{yb}/\gamma_{M0}}{\sigma_{com,Ed}} = 0.64 \times 67.6 \times \frac{280/1.0}{280} = 43.3\,\text{mm}^2$$

$\therefore t_{red} = t A_{s,red}/A_s = 1.56 \times (43.3/67.6) = 1.00\,\text{mm}$

Calculation of effective section properties for distortional buckling

Effective area

$$A_{eff} = (73.87 \times 1.56) + (2 \times 29.16 \times 1.56) + [2 \times (29.16 + 14.2) \times 1.00]$$
$$= 292.8\,\text{mm}^2 \text{ (compared with 341.5 mm}^2 \text{ for local buckling alone)}$$

The horizontal position of the neutral axis from the centreline of the web for the effective section is

$$\bar{y}_{eff} = \{[2 \times (29.16 \times 1.56) \times 29.16/2] + [2 \times (29.16 \times 1.00) \times (63.4 - 29.16/2)]$$
$$+ [2 \times (14.2 \times 1.00) \times 63.4]\}/292.8$$
$$= 20.38\,\text{mm}$$

The horizontal shift in the neutral axis from the gross section to the effective section is

$$e_{Ny} = 20.38 - 16.46 = 3.92 \text{ mm}$$

Step 3: optionally iterate

χ_d may be refined iteratively using modified values of ρ obtained by taking $\sigma_{com,Ed}$ equal to $\chi_d f_{yb}/\gamma_{M0}$ in Step 1 for each iteration. Subsequent steps are as shown in this example.

The compressive resistance of the cross-section, accounting for distortional buckling, is therefore as follows:

$$N_{c,Rd} = A_{eff} f_{yb} = 292.8 \times 280 \times 10^{-3} = 82.0 \text{ kN}$$

There is, however, a shift in the neutral axis of 3.92 mm (from the centroid of the gross section to the centroid of the effective section), and the cross-section should strictly therefore also be checked for combined axial compression plus bending, with the bending moment equal to the applied axial load multiplied by the shift in neutral axis.

Table 13.2. Buckling curve selection table from EN 1993-1-3

Type of cross-section		Buckling about axis	Buckling curve
	If f_{yb} is used	Any	b
	If f_{ya} is used*	Any	c
		y–y z–z	a b
		Any	b
		Any	c

* the average yield strength f_{ya} should not be used unless $A_{eff} = A_g$

Whatever the mode of buckling of a member (i.e. flexural, torsional or torsional–flexural) the generic buckling curve formulations and the method for determining member resistances are common. The only difference is in the calculation of the elastic critical buckling force, which is particular to the mode of buckling, and is used to define $\bar{\lambda}$.

The non-dimensional slenderness $\bar{\lambda}$ is defined by equation (D13.12) for Class 1, 2 and 3 cross-sections, and by equation (D13.13) for Class 4 cross-sections; a subscript T is added to $\bar{\lambda}$ to indicate when the buckling mode includes a torsional component:

$$\bar{\lambda} = \bar{\lambda}_T = \sqrt{\frac{A f_y}{N_{cr}}} \qquad \text{for Class 1, 2 and 3 cross-sections} \tag{D13.12}$$

$$\bar{\lambda} = \bar{\lambda}_T = \sqrt{\frac{A_{eff} f_y}{N_{cr}}} \qquad \text{for Class 4 cross-sections} \tag{D13.13}$$

where, for torsional and torsional–flexural buckling,

$$N_{cr} = N_{cr,TF} \qquad \text{but } N_{cr} \le N_{cr,T}$$

$N_{cr,TF}$ is the elastic critical torsional–flexural buckling force
$N_{cr,T}$ is the elastic critical torsional buckling force.

The elastic critical buckling forces for torsional and torsional–flexural buckling for cross-sections that are symmetrical about the y–y axis (i.e. where $z_0 = 0$) are given by equations (D13.14) and (D13.15), respectively:

$$N_{cr,T} = \frac{1}{i_0^2}\left(GI_t + \frac{\pi^2 EI_w}{l_T^2}\right) \tag{D13.14}$$

where

$$i_0^2 = i_y^2 + i_z^2 + y_0^2 + z_0^2$$

G is the shear modulus
I_t is the torsion constant of the gross cross-section
I_w is the warping constant of the gross cross-section
i_y is the radius of gyration of the gross cross-section about the y–y axis
i_z is the radius of gyration of the gross cross-section about the z–z axis
l_T is the buckling length of the member for torsional buckling
y_0 is the distance from the shear centre to the centroid of the gross cross-section along the y axis
z_0 is the distance from the shear centre to the centroid of the gross cross-section along the z axis.

$$N_{cr,TF} = \frac{N_{cr,y}}{2\beta}\left(1 + \frac{N_{cr,T}}{N_{cr,y}} - \sqrt{\left(1 - \frac{N_{cr,T}}{N_{cr,y}}\right)^2 + 4\left(\frac{y_0}{i_0}\right)^2 \frac{N_{cr,T}}{N_{cr,y}}}\right) \tag{D13.15}$$

where

$$\beta = 1 - \left(\frac{y_0}{i_0}\right)^2$$

$N_{cr,y}$ is the critical force for flexural buckling about the y–y axis

Guidance is provided in EN 1993-1-3 on buckling lengths for components with different degrees of torsional and warping restraint. It is stated that for practical connections at each end, l_T/L_T

Figure 13.13. (a) Partial and (b) significant torsional and warping restraint from practical connections

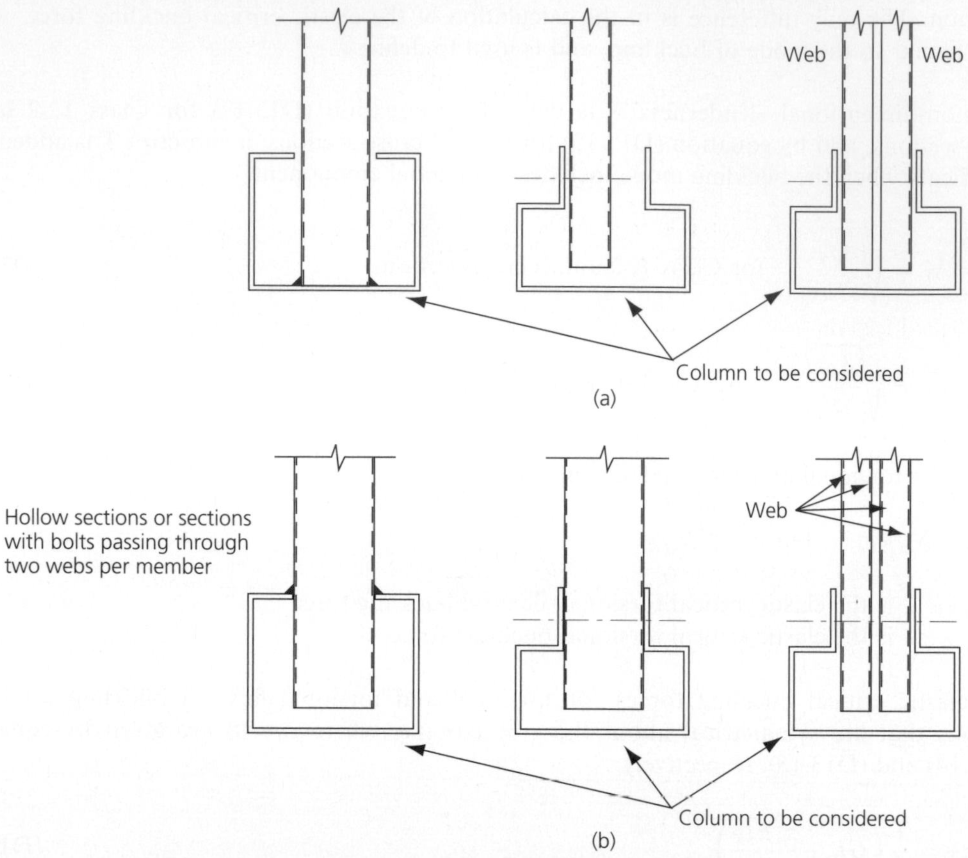

(a)

(b)

(the effective buckling length divided by the system length) should be taken as

- 1.0 for connections that provide partial restraint against torsion and warping (Figure 13.13a)
- 0.7 for connections that provide significant restraint against torsion and warping (Figure 13.13b).

Example 13.3: member resistance in compression (checking flexural, torsional and torsional–flexural buckling)

Calculate the member resistance for a $100 \times 50 \times 3$ plain channel section column subjected to compression. The column length is 1.5 m, with pinned end conditions, so the effective length is assumed to equal to the system length. The steel has a yield strength f_y of 280 N/mm^2, a Young's modulus of 210 000 N/mm^2 and a shear modulus of 81 000 N/mm^2. No allowance will be made for coatings in this example.

From section tables (Gorgolewski *et al.*, 2001), the following details (defined above) are obtained:

$A = 5.55 \, \text{cm}^2$	$i_y = 3.92 \, \text{cm}$	$I_t = 0.1621 \, \text{cm}^4$
$A_{\text{eff}} = 5.49 \, \text{cm}^2$	$i_z = 1.57 \, \text{cm}$	$I_w = 210 \, \text{cm}^6$
$I_y = 85.41 \, \text{cm}^4$	$W_{\text{el,y}} = 17.09 \, \text{cm}^3$	$y_0 = 3.01 \, \text{cm}$
$I_z = 13.76 \, \text{cm}^4$	$W_{\text{el,z}} = 3.83 \, \text{cm}^3$	

Calculate critical buckling loads

Flexural buckling – major (y–y) axis:

$$N_{\mathrm{cr},y} = \frac{\pi^2 E I_y}{L_{\mathrm{cr}}^2} = \frac{\pi^2 \times 210\,000 \times 85.41 \times 10^4}{1500^2} = 787 \times 10^3\,\mathrm{N} = 787\,\mathrm{kN}$$

Flexural buckling – minor (z–z) axis:

$$N_{\mathrm{cr},z} = \frac{\pi^2 E I_z}{L_{\mathrm{cr}}^2} = \frac{\pi^2 \times 210\,000 \times 13.76 \times 10^4}{1500^2} = 127 \times 10^3\,\mathrm{N} = 127\,\mathrm{kN}$$

Torsional buckling:

$$N_{\mathrm{cr,T}} = \frac{1}{i_0^2}\left(GI_t + \frac{\pi^2 E I_w}{l_T^2}\right) \tag{D13.14}$$

$$i_0^2 = i_y^2 + i_z^2 + y_0^2 + z_0^2 = 39.2^2 + 15.7^2 + 30.1^2 = 2689\,\mathrm{mm}^2$$

$$\therefore\ N_{\mathrm{cr,T}} = \frac{1}{2689}\left(81\,000 \times 0.1621 \times 10^4 + \frac{\pi^2 \times 210\,000 \times 210 \times 10^6}{1500^2}\right)$$

$$= 121 \times 10^3\,\mathrm{N} = 121\,\mathrm{kN}$$

Torsional–flexural buckling:

$$N_{\mathrm{cr,TF}} = \frac{N_{\mathrm{cr},y}}{2\beta}\left(1 + \frac{N_{\mathrm{cr,T}}}{N_{\mathrm{cr},y}} - \sqrt{\left(1 - \frac{N_{\mathrm{cr,T}}}{N_{\mathrm{cr},y}}\right)^2 + 4\left(\frac{y_0}{i_0}\right)^2 \frac{N_{\mathrm{cr,T}}}{N_{\mathrm{cr},y}}}\right) \tag{D13.15}$$

$$\beta = 1 - \left(\frac{y_0}{i_0}\right)^2 = 1 - \left(\frac{30.1}{51.9}\right)^2 = 0.66$$

$$N_{\mathrm{cr,TF}} = \frac{787 \times 10^3}{2 \times 0.66}\left(1 + \frac{121 \times 10^3}{787 \times 10^3} - \sqrt{\left(1 - \frac{121 \times 10^3}{787 \times 10^3}\right)^2 + 4 \times \left(\frac{30.1}{51.9}\right)^2 \times \frac{121 \times 10^3}{787 \times 10^3}}\right)$$

$$= 114 \times 10^3\,\mathrm{N} = 114\,\mathrm{kN}$$

Torsional–flexural buckling is critical (with $N_{\mathrm{cr}} = 114\,\mathrm{kN}$).

Member buckling resistance

$$N_{\mathrm{b,Rd}} = \frac{\chi A_{\mathrm{eff}} f_y}{\gamma_{\mathrm{M1}}} \quad \text{for Class 4 cross-sections} \tag{6.47}$$

$$\chi = \frac{1}{\Phi + \sqrt{\Phi^2 - \bar{\lambda}^2}} \quad \text{but } \chi \leq 1.0 \tag{6.49}$$

where

$$\Phi = 0.5[1 + \alpha(\bar{\lambda} - 0.2) + \bar{\lambda}^2]$$

$$\bar{\lambda} = \sqrt{\frac{A_{\mathrm{eff}} f_y}{N_{\mathrm{cr}}}} \quad \text{for Class 4 cross-sections}$$

Non-dimensional slenderness (for torsional–flexural buckling mode)

$$\therefore\ \bar{\lambda} = \sqrt{\frac{549 \times 280}{114 \times 10^3}} = 1.16$$

Selection of buckling curve and imperfection factor α
For cold-formed plain channel sections, use buckling curve c (see Table 13.2).

For buckling curve c, $\alpha = 0.49$ (see Table 6.4 (*Table 6.1* of EN 1993-1-1))

Buckling curves

$$\Phi = 0.5[1 + 0.49 \times (1.16 - 0.2) + 1.16^2] = 1.41$$

$$\chi = \frac{1}{1.41 + \sqrt{1.41^2 - 1.16^2}} = 0.45$$

$$N_{b,Rd} = \frac{0.45 \times 549 \times 280}{1.0} = 69.2 \times 10^3 \, \text{N} = 69.2 \, \text{kN}$$

The member resistance of the $100 \times 50 \times 3$ plain channel (governed by torsional–flexural buckling) is 69.2 kN.

13.8. Shear lag

No guidance on the effects of shear lag is given in EN 1993-1-3, except to say that shear lag shall be taken into account according to EN 1993-1-5. An introduction to the provisions for shear lag of EN 1993-1-5 is given in Section 6.2.2 of this guide.

13.9. Flange curling

EN 1993-1-3 states that the effect of flange curling on the load-bearing resistance should be taken into consideration when the magnitude of curling is greater than 5% of the depth of the cross-section. For initially straight beams, equation (D13.16), which applies to both the compression and tension flanges, with or without stiffeners, is provided in clause 5.4 of EN 1993-1-3. For arched beams, where the curvature, and therefore the transverse force components on the flanges, are larger, equation (D13.17) is provided.

$$u = 2\frac{\sigma_a^2}{E^2}\frac{b_s^4}{t^2 z} \tag{D13.16}$$

$$u = 2\frac{\sigma_a}{E}\frac{b_s^4}{t^2 r} \tag{D13.17}$$

where

u is the magnitude of the flange curling (towards the neutral axis)
b_s is one half of the distance between webs in box or hat sections, or the width of the portion of flange projecting from the web
t is the flange thickness
z is the distance of the flange under consideration to the neutral axis
r is the radius of curvature of an arched beam
σ_a is the mean stress in the flange.

If the stress has been calculated over the effective cross-section, the mean stress is obtained by multiplying the stress for the effective cross-section by the ratio of the effective cross-section by the ratio of the effective flange area to the gross flange area.

If the magnitude of flange curling is found to be greater than 5% of the depth of the cross-section, then a reduction in load-bearing capacity due, for example, to the effective reduction in depth of the section or due to possible bending of the web, should be made.

13.10. Web crushing, crippling and buckling

Transversely loaded webs of slender proportions, which are common in cold-formed sections, are susceptible to a number of possible forms of failure, including web crushing, web crippling and web buckling. Web crushing involves yielding of the web material directly adjacent to the flange. Web crippling describes a form of failure whereby localised buckling of the web beneath the transversely loaded flange is accompanied by web crushing and plastic deformation of the flange. Transversely loaded webs can also fail as a result of overall web buckling, with the web acting as a strut. This form of failure requires that the transverse load is carried from the loaded flange through the web to a reaction at the other flange.

Calculation of the transverse resistance of a web $R_{w,Rd}$ involves categorisation of the cross-section and determination of a number of constants relating to the properties of the cross-section and loading details. Three categories are defined: cross-sections with a single unstiffened web; cross-sections or sheeting with two or more unstiffened webs; and stiffened webs. Web resistance is specified by a number of expressions, selection of which is based principally on the position and nature of loading and reactions, including the proximity of the loading or reactions to free ends.

REFERENCES

Gorgolewski MT, Grubb PJ and Lawson RM (2001) *Modular Construction Using Light Steel Framing*. Steel Construction Institute, Ascot, P302.

Grubb PJ, Gorgolewski MT and Lawson RM (2001) *Building Design Using Cold Formed Steel Sections*. Steel Construction Institute, Ascot, P301.

Rhodes J and Lawson RM (1992) *Design of Structures Using Cold Formed Steel Sections*. Steel Construction Institute, Ascot, P089.

Designers' Guide to Eurocode 3: Design of Steel Buildings, 2nd ed.
ISBN 978-0-7277-4172-1

ICE Publishing: All rights reserved
doi: 10.1680/dsb.41721.151

Chapter 14
Actions and combinations of actions

14.1. Introduction

As noted in Chapter 1 of this guide, EN 1993-1-1 is not a self-contained document for the design of steel structures, but rather provides the rules that are specific to steel structures. Actions (or loads) and combinations of actions, for example, are (other than self-weights) broadly independent of the structural material and are thus contained elsewhere.

This chapter contains a brief review of the guidance provided by EN 1990 and parts of EN 1991 relating to actions and combinations of actions for steel structures.

The basic requirements of EN 1990 state that a structure shall have adequate structural resistance (ultimate limit states), serviceability (serviceability limit states), durability, fire resistance and robustness. For ultimate limit states, checks should be carried out for the following, as relevant:

- EQU – loss of static equilibrium of the structure or any part of the structure
- STR – internal failure or excessive deformation of the structure or structural members
- GEO – failure or excessive deformation of the ground
- FAT – fatigue failure of the structure or structural members.

EN 1990 also emphasises, in Section 3, that all relevant design situations must be examined. It states that

> the selected design situations shall be sufficiently severe and varied so as to encompass all conditions that can reasonably be foreseen to occur during execution and use of the structure.

Design situations are classified as follows:

- persistent design situations, which refer to conditions of normal use
- transient design situations, which refer to temporary conditions, such as during execution or repair
- accidental design situations, which refer to exceptional conditions such as fire, explosion or impact
- seismic design situations, which refer to conditions where the structure is subjected to seismic events.

14.2. Actions

In EN 1990, actions are classified by their variation with time, as permanent, variable or accidental. Permanent actions (G) are those that essentially do not vary with time, such as the self-weight of a structure and fixed equipment; these have generally been referred to as dead loads in British Standards. Variable actions (Q) are those that can vary with time, such as imposed loads, wind loads and snow loads; these have generally been referred to as live loads in British Standards. Accidental actions (A) are usually of short duration, but high magnitude, such as explosions and impacts. Actions should also be classified by their origin, spatial variation and nature, as stated in clause 4.1.1 of EN 1990. Classification by variation with time is important for the establishment of combinations of actions, while the other classifications are necessary for the evaluation of representative values of actions.

Actions on structures may be determined with reference to the appropriate parts of EN 1991, *Eurocode 1: Actions on Structures* and their respective UK National Annexes. EN 1991 contains the following parts:

- Part 1.1, *General Actions – Densities, Self-weight, Imposed Loads for Buildings*
- Part 1.2, *General Actions – Actions on Structures Exposed to Fire*
- Part 1.3, *General Actions – Snow Loads*
- Part 1.4, *General Actions – Wind Actions*
- Part 1.5, *General Actions – Thermal Actions*
- Part 1.6, *General Actions – Actions During Execution*
- Part 1.7, *General Actions – Accidental Actions from Impact and Explosions.*

14.3. Fundamental combinations of actions
14.3.1 General

Clause 6.4.3.2 of EN 1990 provides two options for determining the fundamental combination of actions (load combinations) at ultimate limit states. 'Fundamental' refers to the persistent or transient design situations, rather than accidental or seismic design situations. The first of the options is given by equation (D14.1), which is equation (6.10) of EN 1990. This equation must be used for checking overall equilibrium of the structure as a rigid body (i.e. EQU limit states), and may also be used for STR and GEO limit states.

$$\sum_{j \geq 1} \gamma_{G,j} G_{k,j} \;'+'\; \gamma_P P \;'+'\; \gamma_{Q,1} Q_{k,1} \;'+'\; \sum_{i > 1} \gamma_{Q,i} \psi_{0,i} Q_{k,i} \tag{D14.1}$$

The second option, which may be applied to STR and GEO limit states, is to define load combinations from the less favourable of the two following expressions, which are equations (6.10a) and (6.10b) of EN 1990:

$$\sum_{j \geq 1} \gamma_{G,j} G_{k,j} \;'+'\; \gamma_P P \;'+'\; \gamma_{Q,1} \psi_{0,1} Q_{k,1} \;'+'\; \sum_{i > 1} \gamma_{Q,i} \psi_{0,i} Q_{k,i} \tag{D14.2a}$$

$$\sum_{j \geq 1} \xi_j \gamma_{G,j} G_{k,j} \;'+'\; \gamma_P P \;'+'\; \gamma_{Q,1} Q_{k,1} \;'+'\; \sum_{i > 1} \gamma_{Q,i} \psi_{0,i} Q_{k,i} \tag{D14.2b}$$

where

'+'	implies 'to be combined with'
	implies 'the combined effect of'
ψ_0	is a combination factor, discussed below
ξ	is a reduction factor for unfavourable permanent actions G, discussed below
γ_G	is a partial factor for permanent actions
γ_Q	is a partial factor for variable actions
P	represents actions due to prestressing.

The combination factor ψ_0 is one of three ψ factors (ψ_0, ψ_1 and ψ_2) used in EN 1990. The purpose of the ψ factors is to modify characteristic values of variable actions to give representative values for different situations. The combination factor ψ_0 is intended specifically to take account of the reduced probability of the simultaneous occurrence of two or more variable actions, and hence appears in each of equations (D14.1), (D14.2a) and (D14.2b). ψ factors are discussed in Section 4.1.3 of EN 1990.

ξ appears in equation (D14.2b) (equation (6.10b) of EN 1990), and is a reduction factor for unfavourable permanent actions G. The ξ factor is set in the UK National Annex to EN 1990 equal to 0.925.

Ignoring prestressing actions, which are generally absent in conventional steel structures, each of the combination expressions contains:

- permanent actions $G_{k,j}$
- a leading variable action $Q_{k,1}$
- other variable actions $Q_{k,i}$.

In general, unless it is clearly not a critical combination, each variable action should be considered as the leading variable action, in turn. Clause 6.1(2) of EN 1990 states that actions that cannot occur simultaneously, for example due to physical reasons, should not be considered together in combination.

In any given combination of actions, individual actions may be considered as either 'unfavourable' or 'favourable', depending on whether they cause an increase or reduction in forces and moments in the structural members or are destabilising or stabilising in terms of overall equilibrium. For STR limit states combinations, regarding actions as unfavourable will typically lead to the most severe condition, but this will not always be the case; for example, in situations of wind uplift, the most severe condition may arise when both the permanent and imposed loads are considered to be favourable. For structures sensitive to variation in the magnitude of permanent actions in different parts of the structure, clause 6.4.3.1(4) of EN 1990 states that the upper characteristic value of a permanent action $G_{kj,sup}$ should be used when that action is unfavourable, and the lower characteristic value of a permanent action $G_{kj,inf}$ should be used when that action is favourable. Upper and lower characteristic values of permanent actions may be determined as described in Gulvanessian *et al.* (2002).

For serviceability limit states, guidance on combinations of actions is given in clauses 6.5.3 and A1.4 of EN 1990. Additional information is provided in the UK National Annex to EN 1993-1-1, while further interpretation of the guidance may be found in Chapter 7 of this guide and elsewhere (Gulvanessian *et al.*, 2002).

14.3.2 Buildings

Methods for establishing combinations of actions for buildings are given in Annex A1 of EN 1990. To simplify building design, note 1 to clause A.1.2.1(1) of EN 1990 allows the combination of actions to be based on not more than two variable actions. This simplification is intended only to apply to common cases of building structures.

Recommended values of ψ factors for buildings are given in Table 14.1, which is Table NA.A1.1 from the UK National Annex; this replaces Table A1.1 of EN 1990. For load combinations at ultimate limit state, ψ_0 is the factor of interest. Differences of note between Table NA.A1.1 of the UK National Annex and Table A1.1 of EN 1990 relate to wind and roof loading. For wind loading ψ_0 is given as 0.5 in the UK National Annex compared with 0.6 in EN 1990, while for roof loading a value of 0.7 in the UK National Annex replaces a value of 0.0 in EN 1990.

The design values of actions for ultimate limit states in the persistent and transient design situations are given in Tables 14.2(A)–(C) (Tables NA1.2(A)–(C) of the UK National Annex to EN 1990). Table 14.2(A) provides design values of actions for verifying the static equilibrium (EQU) of building structures. Table 14.2(B) provides design values of actions for the verification of structural members (STR) in buildings, not involving geotechnical action. For the design of structural members (STR) involving geotechnical actions (GEO), three approaches are outlined in clause A1.3.1(5), and reference should be made to Tables 14.2(B) and 14.2(C).

Although, in principle, all possible combinations of actions should be considered, application of equation (6.10) of EN 1990 or equations (6.10a) and (6.10b) together will typically, for regular building structures, reduce to only a few key combinations that are likely to govern for STR ULS design. These are given in Table 14.3 for equation (6.10) and Table 14.4 for equations (6.10a) and (6.10b). Note that, of equations (6.10a) and (6.10b), equation (6.10a) is unlikely to govern unless the permanent actions represent a very high proportion of the total load. Note also, given that the designer has a choice between either equation (6.10) or equations (6.10a) and (6.10b) together for deriving load combinations, the more economic option will generally be sought – use of equations (6.10a) and (6.10b) will always yield more competitive results than equation (6.10), due primarily to the dead load reduction factor ξ in equation (6.10b).

Table 14.1. Recommended values of ψ factors for buildings (Table NA1.1 of the UK National Annex to EN 1990)

Action	ψ_0	ψ_1	ψ_2
Imposed loads in buildings, category (see EN 1991-1-1)			
Category A: domestic, residential areas	0.7	0.5	0.3
Category B: office areas	0.7	0.5	0.3
Category C: congregation areas	0.7	0.7	0.6
Category D: shopping areas	0.7	0.7	0.6
Category E: storage areas	1.0	0.9	0.8
Category F: traffic area, vehicle weight $\leq$30kN	0.7	0.7	0.6
Category G: traffic area, 30kN < vehicle weight $\leq$160kN	0.7	0.5	0.3
Category H: roofs[a]	0.7	0	0
Snow loads on buildings (see EN 1991-1-3)			
For sites located at altitude – H > 1000 m a.s.l.	0.70	0.50	0.20
For sites located at altitude – H $\leq$ 1000 m a.s.l.	0.50	0.20	0
Wind loads on buildings (see EN 1991-1-4)	0.5	0.2	0
Temperature (non-fire) in buildings (see EN 1991-1-5)	0.6	0.5	0
NOTE[a] See also EN 1991-1-1 Clause 3.3.2(1).			

Table 14.2(A). Design values of actions (EQU) (Set A) (Table NA.1.2(a) of the UK National Annex to EN 1990)

Persistent and transient design situations	Permanent actions		Leading variable action[a]	Accompanying variable actions	
	Unfavourable	Favourable		Main (if any)	Others
(Eq. 6.10)	$1.10G_{kj,inf}$	$0.90\ G_{kj,inf}$	$1.5Q_{k,1}$ (0 when favourable)		$1.5\psi_{0,i}Q_{k,i}$ (0 when favourable)

[a] Variable actions are those considered in Table NA.A1.1.

In cases where the verification of static equilibrium also involves the resistance of structural members, as an alternative to two separate verifications based on Tables NA.A.1.2 (A) and A1.2 (B), a combined verification, based on Table NA.A1.2 (A), should be adopted, with the following set of values.

$\gamma_{Gj,sup} = 1.35$

$\gamma_{Gj,inf} = 1.15$

$\gamma_{Q,1} = 1.50$ where unfavourable (0 where favourable)

$\gamma_{Q,i} = 1.50$ where unfavourable (0 where favourable)

provided that applying $\gamma_{Gj,inf} = 1.00$ both to the favourable part and to the unfavourable part of permanent actions does not give a more unfavourable effect.

Table 14.2(B). Design values of actions (STR/GEO) (Set B) (Table NA.1.2(B) of the UK National Annex to EN 1990)

Persistent and transient design situations	Permanent actions		Leading variable action[a]	Accompanying variable actions[a]	
	Unfavourable	Favourable		Main (if any)	Others
(Eq. 6.10)	$1.35G_{kj,sup}$	$1.00G_{kj,inf}$	$1.5Q_{k,1}$		$1.5\psi_{0,i}Q_{k,i}$

Persistent and transient design situations	Permanent actions		Leading variable action[a]	Accompanying variable actions[a]	
	Unfavourable	Favourable	Action	Main	Others
(Eq. 6.10a)	$1.35G_{kj,sup}$	$1.00G_{kj,inf}$		$1.5\psi_{0,1}Q_{k,1}$	$1.5\psi_{0,i}Q_{k,i}$
(Eq. 6.10b)	$0.925*1.35G_{kj,sup}$	$1.00G_{kj,inf}$	$1.5Q_{k,1}$		$1.5\psi_{0,i}Q_{k,i}$

NOTE 1 Either expression 6.10, or expression 6.10a together with 6.10b may be made, as desired.

NOTE 2 The characteristic values of all permanent actions from one source are multiplied by $\gamma_{G,sup}$ if the total resulting action effect is unfavourable and $\gamma_{G,inf}$ if the total resulting action effect is favourable. For example, all actions originating from the self-weight of the structure may be considered as coming from one source; this also applies if different materials are involved.

NOTE 3 For particular verifications, the values for γ_G and γ_Q may be subdivided into γ_g and γ_q and the model uncertainty factor γ_{Sd}. A value of γ_{Sd} in the range 1.05 to 1.15 can be used in most common cases and can be modified in the National Annex.

NOTE 2 When variable actions are favourable Q_k should be taken as 0.

[a] Variable actions are those considered in Table NA.A1.1.

Table 14.2(C). Design values of actions (STR/GEO) (Set C) (Table NA1.2(C) of the UK National Annex to EN 1990)

Persistent and transient design situations	Permanent actions		Leading variable action*	Accompanying variable actions[a]	
	Unfavourable	Favourable		Main (if any)	Others
(Eq. 6.10)	$1.0G_{kj,sup}$	$1.0G_{kj,inf}$	$1.3Q_{k,1}$ (0 where favourable)		$1.3\psi_{0,i}Q_{k,i}$ (0 where favourable)

[a] Variable actions are those considered in Table NA.A1.1.

Table 14.3. Typical STR combinations of actions arising from equation (6.10) of EN 1990

Combination	Load factor γ		
	Permanent γ_G	Imposed γ_Q	Wind γ_Q
Permanent + imposed	1.35	1.5	–
Permanent + wind (uplift)	1.0	–	1.5
Permanent + imposed + wind (imposed leading)	1.35	1.5	0.75
Permanent + imposed + wind (wind leading)	1.35	1.05	1.5

Table 14.4. Typical STR combinations of actions arising from equations (6.10a) and (6.10b) of EN 1990

Combination	Load factor γ		
	Dead γ_G	Imposed γ_Q	Wind γ_Q
Permanent + imposed: equation (6.10a)	1.35	1.05	–
Permanent + imposed + wind: equation (6.10a)	1.35	1.05	0.75
Permanent + imposed: equation (6.10b)	1.25	1.5	–
Permanent + wind (uplift): equation (6.10b)	1.0	–	1.5
Permanent + imposed + wind (imposed leading): equation (6.10b)	1.25	1.5	0.75
Permanent + imposed + wind (wind leading): equation (6.10b)	1.25	0.75	1.5

REFERENCE

Gulvanessian H, Calgaro J-A and Holický M (2002) *Designers' Guide to EN 1990, Eurocode: Basis of Structural Design.* Thomas Telford, London.

Designers' Guide to Eurocode 3: Design of Steel Buildings, 2nd ed.
ISBN 978-0-7277-4172-1

ICE Publishing: All rights reserved
doi: 10.1680/dsb.41721.157

Index

Page locators in *italics* refer to figures separate from the corresponding text.